December 1996

British Bus Publishing

The Fire Brigade Handbook

The Fire Brigade Handbook is a directory of the fire appliances used by the various fire authorities in England, Scotland and Wales. The form describes the fleet number, registration and type, together with the most recent station to which it was allocated. Invariably appliances move from time to time to meet the demands placed on the brigades.

This is the second edition of our Fire Brigade Handbook listing all the appliances of fire and rescue brigades based in England, Scotland and Wales, and follows similar lists of bus and coach operators, also available from the publisher. It is hoped to build on the information and photographs in the book and, with that in mind, the publisher would like to hear from brigades, suppliers and individuals who can supply accurate information and quality pictures for the next edition.

The publishers unfortunately cannot accept responsibility for any loss and request you show your name on each picture or slide.

Principal Editor: Keith Grimes.

Acknowledgements:
We are grateful to Angloco Ltd, Graham Brown, Andrew Daley, Robert Hawkes, Alan Hewitt, Jef Johnson, Steve Jones, Colin Lloyd, Alistair MacDonald, David Mitchell, Clive Shearman, Robert Smith, Gavin Stewart, Karl Sillitoe, Peter Wilson and the Management and officials of the various authorities who have assisted in the preparartion of this publication, especially those involved in the recent county reorganisation.

ISBN 1 897990 52 9
Published by *British Bus Publishing*
The Vyne, 16 St Margarets Drive, Wellington,
Telford, Shropshire, TF1 3PH
© British Bus Publishing, December 1996

CONTENTS

Introductory Notes on fire appliance designations.

All fire brigades designate different types of fire appliance according to their use and the equipment carried. The list below is an index to those codes used in the book and modified where similar abbrviations are used between brigades for differing needs - such as LRU for Light Rescue Unit and Line Rescue Unit. Here we have ensured consistancy by using LiRU for the latter.

Several fire brigades change the class of vehicle where extra equipment is carries, such as heavy cutting equipment, or when ladders are transferred between pumps to meet the local needs. Other vehicles perform a dual role and thus carry two sets of codes.

ALP	Aerial Ladder Platform	L4P	Light four-wheel drive pump
AP	Aerial Platform	L4PM	Light four-wheel drive Prime Mover
APL	Aerial Platform with ladder	L4RP	Light four-wheel drive Rescue Pump
ATV	All Terrain Vehicle	L4T	Light four-wheel drive pump and hose
BA(T)	Breathing Apparatus Training Unit	L6P	Light six-wheel pump
BAMU	Breathing Apparatus Maintenance Unit	L6V	Light six-wheel vehicle
BASU	BA Support Unit	L8V	Light eight-wheel drive vehicle
BAT	Breathing Apparatus Tender	LFA	Light fire appliance
BTU	Boat & Trailer Unit	LR	Land Rover
BWrC	Bulk Water Carrier	LRU	Light Rescue Unit
C/FRT	Compact Fire Rescue Tender	LiRU	Line Rescue Unit
CaV	Canteen Unit	MRT	Major Rescue Tender
CEU	Community Education Unit	MRV	Major Recovery Vehicle
CIU	Chemical Incident Unit	MWrT	Mini Water Tender
CP	Compact Pump	OSU	Operations Support Unit
CRU	Cliff Rescue Unit	P	Pump
CT	Crash Tender	PCV	Personnel Carrier
CU	Control Unit	PE	Pump with wheeled escape ladder
DCU	Damage Control Unit	PHP	Pump with Hydraulic Platform.
DecU	Decontamination Unit	PL	Pump ladder (13.5m)
DLU	Damage Limitation Unit	PM	Prime Mover
DTV	Driver Training Vehicle	PRL	Pump/Rescue Tender Ladder
EBt	Estuary Boat	RDCU	Rescue/Damage Control Unit
EPU	Environment Protection Unit	RIP	Rapid Intervention Pump
ERT	Emergency Rescue Tender	RP	Rescue Pump
EST	Emergency Salvage Tender	RRRT	Rapid Response Rescue Tender
ET	Emergency Tender	RSU	Rail Support Unit
ExU	Exhibition Unit	RSV	Rescue Support Vehicle
FBL	Flat-bed lorry	RT	Rescue Tender
FBt	Fire Boat - non inflatable	RU	Rescue Unit
FIU	Fire Investigation Unit	RV	Recovery Vehicle
FLT	Folk Lift Truck	RWrL	Rescue/Water Tender Ladder
FoT	Foam Tender	S4x4	Small 4x4 Pump
FRT	Fire Rescue Tender	SEU	Special Equipment Unit
FST	Foam Salvage Tender	SIU	Special Incident Unit
GPV	General Purpose Vehicle	SRP	Snozzle Rescue Pump
H4P	Heavy 4-wheel drive Pump	SRU	Special Rescue Unit
HL	Hose Layer	SSU	Special Support Unit
HMU	Hazardous Materials Unit	ST	Salvage Tender
HP	Hydraulic Platform	SU	Support Unit
HRU	Heavy Rescue Unit	SU	Salvage Unit
HSRU	Hazardous Substances Rescue Unit	TL	Turntable Ladder
ICCU	Incident Command & Control Unit	TU	Towing Unit
ICU	Incident Command Unit	ULFA	Ultra Light Fire Appliance
IRBt	Inflatable Rescue Boat	UV	Uniform Van
ISU	Insident Support Unit	WrC	Water Carrier
IU	Information Unit	WrE	Water tender Escape
IWBt	Inland Water Boat	WrL	Water tender with 13.5m ladder
LET	Light Emergency Tender	WrT	Water tender

AVON FIRE BRIGADE

Avon Fire Brigade, Temple Back, Bristol BS1 6EU

A/01/93	L411SHT	Mitsubishi L200 4x4	Mitsubishi Pick-up	L4V	1993	A2 Southmead, Bristol
A/08/93	L412SHT	Mitsubishi L200 4x4	Mitsubishi Pick-up	L4V	1993	B1 Bath
A/10/93	L413SHT	Mitsubishi L200 4x4	Mitsubishi Pick-up	LiRU	1993	C1 Weston-super-Mare
A/29/89	F58RTC	Renault-Dodge G11	Avon FB	GPV	1989	Training Centre
A/30/89	F59RTC	Renault-Dodge G11	Avon FB	GPV	1989	Training Centre
F/01/89	F65RTC	Renault-Dodge G13	Saxon Sanbec	WrL	1989	Driving School
F/02/91	J639GHY	Renault Midliner M230-15D	Saxon Sanbec	WrL	1991	C10 Yatton
F/03/92	K748LWS	Renault Midliner M230-15D	Saxon Sanbec	WrT	1992	A7 Yate
F/04/92	K749LWS	Renault Midliner M230-15D	Saxon Sanbec	WrL	1992	B7 Kingswood, Bristol
F/05/92	K750LWS	Renault Midliner M230-15D	Saxon Sanbec	WrL	1992	B6 Speedwell, Bristol
F/06/93	L547SAE	Renault Midliner M230-15D	Saxon Sanbec	WrL	1993	A5 Patchway
F/07/93	L548SAE	Renault Midliner M230-15D	Saxon Sanbec	WrL	1993	A3 Avonmouth
F/08/92	K751LWS	Renault Midliner M230-15D	Saxon Sanbec	WrL	1992	C5 Bedminster, Bristol
F/09/93	L545SAE	Renault Midliner M230-15D	Saxon Sanbec	WrT	1993	C1 Weston-super-Mare
F/10/91	J642GHY	Renault Midliner M230-15D	Saxon Sanbec	WrT	1991	C9 Nailsea
F/12/92	K752LWS	Renault Midliner M230-15D	Saxon Sanbec	WrL	1992	A2 Southmead, Bristol
F/13/91	J643GHY	Renault Midliner M230-15D	Saxon Sanbec	WrT	1991	A3 Avonmouth
F/14/93	L546SAE	Renault Midliner M230-15D	Saxon Sanbec	WrL	1993	A7 Yate
F/15/94	M206AOU	Renault Midliner M230-15D	Saxon Sanbec	WrT	1994	C5 Bedminster, Bristol
F/15/80	DHY523W	Dodge G1313	HCB-Angus	WrL	1980	On loan to British Aerospace
F/16/83	NEU560Y	Dodge G1313	Merryweather	WrT	1983	Training Centre
F/17/94	M207AOU	Renault Midliner M230-15D	Saxon Sanbec	WrL	1994	C1 Weston-super-Mare
F/18/84	B442VAE	Dodge G13	Saxon Sanbec	WrL	1984	B Reserve
F/20/95	M479EAE	Renault Midliner M230-15D	Saxon Sanbec	WrL	1995	B1 Bath
F/21/95	M480EAE	Renault Midliner M230-15D	Saxon Sanbec	WrL	1995	B1 Bath
F/22/94	M208AOU	Renault Midliner M230-15D	Saxon Sanbec	WrT	1994	B6 Speedwell, Bristol
F/23/96	N602MHY	Renault Midliner M230-15D	Saxon Sanbec	ADTU	1996	Training Centre

The origins of the Merryweather company date back to 1692 from when they produced fire fighting equipment until they ceased trading in the 1980s. Among the last fire appliances to be built were a batch of WrL appliances on Dodge G1313 chassis delivered to Avon in 1983. Two survivors are now allocated to the Training School where F/24/83, NEU562Y, was photographed. *Keith Grimes*

F/24/83	NEU562Y	Dodge G1313	Merryweather	WrL	1983	Training Centre
F/25/95	M277DOU	Renault Midliner M230-15D	Saxon Sanbec	WrT	1995	A1 Temple Back, Bristol
F/26/95	N842HFB	Renault Maxter G330-26D	A G Bracey/Zweiweg	RSU	1995	A3 Avonmouth
F/27/95	M278DOU	Renault Midliner M230-15D	Saxon Sanbec	WrL	1995	A1 Temple Back, Bristol
F/28/84	B443VAE	Dodge G13	Saxon Sanbec	WrL	1984	B Reserve
F/29/84	B444VAE	Dodge G13	Saxon Sanbec	WrL	1984	A Reserve
F/30/95	M279DOU	Renault Midliner M230-15D	Saxon Sanbec	WrT	1995	A1 Temple Back, Bristol
F/31/84	B445VAE	Dodge G13	Saxon Sanbec	WrL	1984	B Reserve
F/32/94	M209AOU	Renault Midliner M230-15D	Saxon Sanbec	WrL	1994	B4 Brislington, Bristol
F/33/84	B446VAE	Dodge G13	Saxon Sanbec	WrL	1984	A Reserve
F/34/84	B447VAE	Dodge G13	Saxon Sanbec	WrL	1984	B Reserve
F/35/85	B448VAE	Dodge G13	Saxon Sanbec	WrL	1985	C Reserve
F/36/86	B169XHY	Land Rover 110 Tdi	Avon CFB	L4V	1985	A1 Temple Back, Bristol
F/37/86	C124BHY	Renault-Dodge G13C	Saxon Sanbec	WrT	1986	A Reserve
F/39/87	E651JOU	Renault-Dodge G13	Saxon Sanbec	WrL	1987	A6 Thornbury
F/40/87	E652JOU	Renault-Dodge G13	Saxon Sanbec	WrT	1987	A6 Thornbury
F/41/87	E653JOU	Renault-Dodge G13	Saxon Sanbec	WrL	1987	C6 Chew Magna
F/42/88	F831OHW	Renault-Dodge G13	Saxon Sanbec	WrL	1988	B2 Radstock
F/43/88	F832OHW	Renault-Dodge G13	Saxon Sanbec	WrT	1988	C4 Pill
F/44/88	F833OHW	Renault-Dodge G13	Saxon Sanbec	WrL	1988	C3 Portishead
F/45/88	F834OHW	Renault-Dodge G13	Saxon Sanbec	WrT	1988	C3 Portishead
F/46/89	F63RTC	Renault-Dodge G13	Saxon Sanbec	WrL	1989	C2 Clevedon
F/47/89	F64RTC	Renault-Dodge G13	Saxon Sanbec	WrT	1989	C8 Winscombe
F/48/91	J640GHY	Renault Midliner M230-15D	Saxon Sanbec	WrL	1991	C9 Nailsea
F/49/89	F66RTC	Renault-Dodge G13	Saxon Sanbec	WrL	1989	B1 Bath
F/50/89	F67RTC	Renault-Dodge G13	Saxon Sanbec	WrL	1989	C7 Blagdon
F/51/91	J641GHY	Renault Midliner M230-15D	Saxon Sanbec	WrL	1991	B5 Keynsham
F/52/89	F848SHW	Renault-Dodge G13	Saxon Sanbec	WrL	1989	C1 Weston-super-Mare
F/53/89	F849SHW	Renault-Dodge G13	Saxon Sanbec	WrT	1989	B3 Paulton
F/54/89	F847SHW	Renault-Dodge G13	Saxon Sanbec	WrT	1989	C2 Clevedon
F/55/95	M210AOU	Mercedes-Benz 1524	Angloco/Metz DLK30PLC	TL	1995	C1 Weston-super-Mare
F/56/86	D951ETC	Iveco-Magirus 256D14	Carmichael / Magirus 23-12	TL	1986	B1 Bath
F/57/86	D950ETC	Iveco-Magirus 256D14	Carmichael / Magirus 23-12	TL	1986	C5 Bedminster
F/58/96	N601LHT	Mercedes-Benz 1524	Angloco/Metz DLK30PLC	TL	1996	A1 Temple Back, Bristol
F/59/80	DHY531W	Dodge G1613	Carmichael/Simon Snorkel SS220	HP	1980	B6 Speedwell, Bristol
F/60/89	F62RTC	Scania 93M-280	Carmichael/Bronto Skylift 22.2Ti	ALP	1989	A3 Avonmouth
F/61/92	K753LWS	Mercedes-Benz 917AF	Saxon Sanbec	RT	1992	C1 Weston-super-Mare
F/62/89	F835OHW	Mercedes-Benz 917AF	Fulton & Wylie	RT	1989	B1 Bath
F/63/96	N603MHY	Renault Midliner M230-15D	Saxon Sanbec	RT	1996	A1 Temple Back, Bristol
F/64/90	G522VWS	Scania 113M-340	Wreckers Int 3000/HAP970	MRT	1990	A5 Patchway
F/65/91	J644GHY	Renault Midliner G300-24D	W.H.Bence	OSU	1991	A3 Avonmouth
F/66/85	AHU388V	Dodge Commando G13	Avon FB/HIAB(1994)	GPV	1985	A3 Avonmouth
F/67/84	B441VAE	Dodge G13C	Carmichael	RT	1984	A Reserve
F/68/85	B828XHY	Dodge S56	Spectra	BASV	1985	A1 Temple Back, Bristol
F/70/88	E668JOU	Land Rover 110 Tdi	Land Rover HCPU/Avon FB	L4V	1988	C5 Bedminster, Bristol
F/71/88	E669JOU	Land Rover 110 Tdi	Land Rover HCPU	L4V	1988	Workshops
F/73/86	C857DEU	Dodge S56	Spectra	CU	1986	B7 Kingswood, Bristol
F/74/86	D952ETC	Renault-Dodge G13	Saxon Sanbec	CIU	1986	B4 Brislington, Bristol
		Moffett Mounty M2003	*Normally carried by J644GHY*	FLT	1991	A3 Avonmouth
		Moffett Mounty M2703	*Normally carried by N842HFB*	FLT	1996	A3 Avonmouth
		Alumi Kart	Railway Trolleys	PCV	1996	A3 Avonmouth
T/1	Trailer	Ifor Williams Trailers	Hose Layer	HL	19	A2 Southmead, Bristol
T/2	Trailer	Ifor Williams Trailers	Hose Layer	HL	19	C1 Weston-super-Mare
T/3	Trailer	Ifor Williams Trailers	Special Rescue Support	SRSU	19	A1 Temple Back, Bristol
Boat	Trailer	Avon Inflatable Supersport S3.40 RIB		IRBt	1992	B1 Bath
Boat	Trailer	Valiant Inflatable V400		IRBt	1995	C5 Bedminster, Bristol

Notes:

F/23/96 can also be used as a Water Carrier
F/66/85 was converted from a Foam Tender
The Alumi Karts are used to transport personnel at rail incidents and used in conjunction with F/26/95 and carried by F/66/85 or F/26/95

Opposite, top: **For several years the preferred intake of pump appliances for Avon were Renault-Dodge G13s with crew-cab bodywork coachbuilt by Saxon Sanbec. Representing the type is F831OHW, an immaculate example based at Radstock.** *Keith Grimes*
Opposite, bottom: **A unique appliance for rail support duties was jointly funded by Avon and Railtrack. This appliance is designed to attend incidents in the nearby Severn Tunnel. Built on a Renault Master chassis, N842HFB features rail wheels by Zweiweg that enable it to run on standard-gauge rails.** *Keith Grimes*

BEDFORDSHIRE FIRE AND RESCUE SERVICE

Bedfordshire Fire and Rescue Service , Southfields Road, Kempston,
Bedford MK42 7NR

1	B21VTM	Land Rover 110	Pilcher Greene	L4P	1984	07 Sandy
3	E933YBH	GMC Custom Deluxe KC30 4x4	HCB-Angus	RU	1987	12 Kempston
4	F146MTM	Land Rover 110	Saxon Sanbec	L4P	1989	09 Woburn
5	E934YBH	GMC Custom Deluxe KC30 4x4	HCB-Angus	RU	1987	Headquarters (R)
7	G727UNK	Scania 93M-280	Angloco / Bronto Skylift 28.2Ti	HP	1990	00 Luton
11	H906HUR	Land Rover 127	HCB-Angus	L4P	1991	06 Potton
14	L547FTM	Scania P93M-280	Angloco / Bronto Skylift 28.2TI	HP	1993	01 Bedford
15	J824MFC	Volvo FL7	Saxon Sanbec	WrC	1992	12 Kempston
16	J703RRO	Volvo FL7	Saxon Sanbec	WrC	1992	10 Toddington
18	H646VNV	Renault-Dodge S66	Ray Smith	PM	1990	03 Leighton Buzzard
21	J623VEG	Iveco TurboDaily 40.10 4x4	Saxon Sanbec	MRV	1992	05 Biggleswade
22	K831DNH	Iveco TurboDaily 40.10 4x4	Saxon Sanbec	MRV	1994	08 Shefford
23	L805SRP	Iveco TurboDaily 40.10 4x4	Saxon Sanbec	MRV	1995	11 Harrold
24	Boat	Compass Inflatable	Rescue Boat	IRBt	1994	01 Bedford
25	M644EVV	Iveco TurboDaily 40.10 4x4	Saxon Sanbec	MRV	1995	04 Ampthill
26	N539ENK	Iveco TurboDaily 40.10 4x4	Saxon Sanbec	MRV	1996	01 Bedford
27	L351HNV	Volvo FL6.18	Bedwas	ERU	1994	13 Stopsley
28	L607KUD	Volvo FL6.18	Bedwas	ERU	1994	01 Bedford
30	L689GVS	Iveco-Ford Eurocargo 75E15	G C Smith	ICU	1994	10 Toddington
43	Pod	Ray Palmer	Foam Unit	FoU	1991	03 Leighton Buzzard
44	Pod	Ray Palmer	Foam Unit	FoU	1993	03 Leighton Buzzard
65	G752UBW	Volvo FL6.14	Saxon Sanbec	WrL	1990	Driver Training
66	D21NGS	Dennis SS135	Angloco	WrL	1986	01 Bedford (R)
67	D22NGS	Dennis SS135	Angloco	WrL	1986	06 Potton
68	B38YNM	Dennis SS135	Dennis	WrL	1985	Training Centre
70	D70RNK	Dennis SS135	Angloco	RP	1987	Headquarters (R)
71	D71RNK	Dennis SS135	Angloco	WrL	1987	04 Ampthill
72	D72RNK	Dennis SS135	Angloco	WrL	1987	Luton (R)
73	E75CKX	Dennis SS135	Carmichael	RP	1988	11 Harrold
74	E74CKX	Dennis SS135	Carmichael	WrL	1988	05 Biggleswade
75	F528NWL	Volvo FL6.14	HCB-Angus	WrL	1989	10 Toddington
76	F529NWL	Volvo FL6.14	HCB-Angus	WrL	1989	01 Bedford
77	J619RRO	Volvo FL6.14	Excalibur CBK	RP	1992	07 Sandy
78	F530NWL	Volvo FL6.14	HCB-Angus	WrL	1989	12 Kempston
79	G801TBW	Volvo FL6.14	Saxon Sanbec	RP	1990	00 Luton
80	J620RRO	Volvo FL6.14	Excalibur CBK	WrL	1992	03 Leighton Buzzard
81	H417FUD	Volvo FL6.14	Excalibur CBK	WrL	1991	Woburn
82	G802TBW	Volvo FL6.14	Saxon Sanbec	RP	1990	05 Biggleswade
83	G803TBW	Volvo FL6.14	Saxon Sanbec	RP	1990	04 Ampthill
84	H418FUD	Volvo FL6.14	Excalibur CBK	RP	1991	08 Shefford
85	K173YWL	Volvo FL6.14	Excalibur CBK	WrL	1993	02 Dunstable
86	K174YWL	Volvo FL6.14	Excalibur CBK	RP	1993	13 Stopsley
87	DKX708T	Dennis R133	Dennis	WrL	1978	Fire Safety
88	N409CBW	Volvo FL6.14	Excalibur CBK	RP	1995	01 Bedford
89	N408CBW	Volvo FL6.14	Excalibur CBK	RP	1995	00 Luton
90	N411CBW	Volvo FL6.14	Excalibur CBK	RP	1995	12 Kempston
91	N410CBW	Volvo FL6.14	Excalibur CBK	RP	1995	03 Leighton Buzard
95	KBM899Y	Dennis SS133	Dennis	WrL	1983	FSNBF
96	L604KUD	Volvo FL6.14	Excalibur CBK	RP	1994	02 Dunstable
97	L605KUD	Volvo FL6.14	Excalibur CBK	WrL	1994	00 Luton
98	L606KUD	Volvo FL6.14	Excalibur CBK	RP	1994	10 Toddington
101	K461CGS	Ford Transit VE6	Ford	PCV	1993	Headquarters
121	G663KNV	Ford Transit VE6	Ford	PCV	1993	Headquarters
126	G127AVS	Renault-Dodge G13	Hilsons	GPV	1990	Driving School

The Fire Brigade Handbook

A programme to replace the Land Rover L4Ps in Bedfordshire is partially complete, although the trio that remain on the run are comparatively youthful models. The replacement vehicles are Iveco Turbo Daily 4x4s designated as Multi-Role vehicles. Show here is an L4P at Sandy (No 1 - B21VTM) an example bodied by Pilcher Greene. *Karl Sillitoe*

Bedfordshires pair of Water Carriers are Saxon-bodied Volvo FL7s. They have the ability to pump water while in motion, a valuable facility at extensive grass or scrub fires. Number 16 (J703RRO) is the unit based at Toddington. *Edmund Gray*

ROYAL BERKSHIRE FIRE & RESCUE SERVICE

Royal Berkshire Fire & Rescue Service, 103 Dee Road,
Tilehurst, Reading RG3 4BW

A2	D797WCF	Dennis RS135	Locomotors	WrL	1987	Reserve
A3	H876NGM	Dennis RS237	Reynolds Boughton	WrL	1991	B8 Sonning
A4	E934EDP	Dennis RS137	Locomotors	WrL	1988	Reserve
A5	C154SAN	Dennis RS135	Locomotors	WrL	1986	A6 Lambourn
A6	F501NJM	Dennis RS237	Locomotors	WrL	1989	A5 Hungerford
A8	F503NJM	Dennis RS237	Locomotors	WrL	1989	Reserve
A9	F502NJM	Dennis RS237	Locomotors	WrL	1989	B16 Bracknell
A10	H877NGM	Dennis RS237	Reynolds Boughton	WrL	1991	B9 Wargrave
A11	K161ECF	Dennis RS237	Carmichael	WrL	1994	C19 Maidenhead
A12	B932KDP	Land Rover 127	Carmichael	L4P	1985	A4 Newbury
A15	G878DRX	Dennis RS237	Reynolds Boughton	WrL	1990	C12 Cookham
A16	G877DRX	Dennis RS237	Reynolds Boughton	WrL	1990	Reserve
A18	G876DRX	Dennis RS237	Reynolds Boughton	WrL	1990	C14 Ascot
A19	E935EDP	Dennis RS137	Locomotors	WrL	1988	A7 Pangbourne
A20	D798WCF	Dennis RS135	Locomotors	WrL	1987	Reserve
A21	E936EDP	Dennis RS137	Locomotors	WrL	1988	A11 Mortimer
A22	C153SAN	Dennis RS135	Locomotors	WrL	1986	B15 Crowthorne
A23	J948VCF	Dennis RS237	Reynolds Boughton	WrL	1993	A4 Newbury
A30	A678AGM	Dennis RS133	Dennis	WrL	1984	Driving School
A34	B532GJB	Dennis RS133	Dennis	WrL	1985	Training Centre

Newer of a pair of Hydraulic platforms for Royal Berkshire is S5 (F961RMO), a Mercedes-Benz 1625 heavy duty chassis with Saxon bodywork. The latter style of Simon Snorkel SS263 booms fitted to this vehicle feature the revised style of jacks that enable easier pitching of the appliance in confined or constricted locations. *Clive Shearman*

An unusual pair of appliances are on the run with the Royal Berkshire Fire & Rescue Service. These have MAN 11.190 single-deck bus chassis and bodywork by Leicester Carriage Company and display panels featuring common bus styling. S15 (K693YMO) is a Chemical Incident Unit at Whitley Wood fire station. *Clive Shearman*

A37	J949VCF	Dennis RS237	Reynolds Boughton	WrL	1993	C19 Maidenhead
A38	K163ECF	Dennis RS237	Carmichael	WrL	1994	C13 Windsor
A39	K162ECF	Dennis RS237	Carmichael	WrL	1994	C17 Slough
A41	D799WCF	Dennis RS135	Locomotors	WrL	1987	B10 Wokingham
A42	K641DJB	Dennis RS237	Carmichael	WrL	1994	A3 Dee Road, Reading
A43	K642DJB	Dennis RS237	Carmichael	WrL	1994	B2 Wokingham Road, Reading
A44	K643DJB	Dennis RS237	Carmichael	WrL	1994	B16 Bracknell
A45	L640LMO	Dennis RS237	Carmichael International	WrL	1995	A1 Caversham Road, Reading
A46	L641LMO	Dennis RS237	Carmichael International	WrL	1995	C17 Slough
A47	N964WRD	Dennis Sabre TSD233	John Dennis	RP	1996	A4 Newbury
A48	N965WRD	Dennis Sabre TSD233	John Dennis	RP	1996	B20 Whitley Wood, Reading
A49	N966WRD	Dennis Sabre TSD233	John Dennis	RP	1996	C18 Langley
S1	K644DJB	Mercedes-Benz 1124AF	Locomotors/Palfinger	RSV	1993	A4 Newbury
S2	Fire Alpha	Rigid Freezer	Boat & Trailer	FBt	1988	A1 Caversham Road, Reading
S3	C206TDP	GMC KC30 4x4	Locomotors	RSV	1986	Reserve
S5	F961RMO	Mercedes-Benz 1625	Saxon / Simon Snorkel SS263	HP	1989	B20 Whitley Wood, Reading
S6	E403DBL	Renault-Dodge 50S46	Locomotors	CaV	1988	B10 Wokingham
S7	Trailer	Tow-a-Van	B A Support	BAS	1985	A7 Pangbourne
S8	Trailer	Tow-a-Van	B A Support	BAS	1985	C19 Maidenhead
S9	WDP169Y	Shelvoke & Drewry WY	Angloco / Simon Snorkel SS263	HP	1983	C17 Slough
S12	F143FWL	Volvo FL6-17	HCB-Angus	PM	1989	B20 Whitley Wood, Reading
S14	F291OGM	GMC KC30 4x4	Excalibur CBK	RSV	1989	C17 Slough
S15	K693YMO	MAN 11.190	Leicester Carriage	CIU	1994	B20 Whitley Wood, Reading
S16	L508KAN	MAN 11.190	Leicester Carriage	ICU	1994	A3 Dee Road, Reading
V5	F194RUU	Land Rover 110	Land Rover/RBFRS	TU	1989	A11 Mortimer
V14	G700CRD	Land Rover 110	Locomotors	TU	1990	C19 Maidenhead
V18	K995BJH	Land Rover Defender 110	Land Rover/RBFRS	TU	1990	A1 Caversham Road, Reading
V20	G776DRU	Land Rover 110	Locomotors	TU	1990	A7 Pangbourne
V24	G697CRD	Land Rover 110	Locomotors	TU	1990	A6 Lambourne
	Pod	Fergussons	Foam Tender	FoT		B20 Whitley Wood, Reading
	Pod	Wilsdon	Damage Control	DCU		B20 Whitley Wood, Reading
	Pod	Fergussons	Water Carrier	WrC		B20 Whitley Wood, Reading

BUCKINGHAMSHIRE FIRE & RESCUE SERVICE

Buckinghamshire Fire & Rescue Service, Cambridge Street,
Aylesbury, Buckinghamshire HP20 1BD

CVV546X	Dodge G1313	Cheshire Fire Engineering	WrT	1981	Buckingham
EMJ300Y	Bedford TM 4x4	Wreckers International Dominator	RV	1982	Workshops
A283NRO	Dodge G13C	Saxon Sanbec	WrT	1983	Winslow
A285NRO	Dodge G13C	Saxon Sanbec	WrT	1983	Olney
A286NRO	Dodge G13C	Saxon Sanbec	WrT	1983	Reserve
B376VWL	Dodge G13C	Saxon Sanbec	WrT	1984	Aylesbury
B377VWL	Dodge G13C	Saxon Sanbec	WrT	1984	Haddenham
B378VWL	Dodge G13C	Saxon Sanbec	WrT	1984	Waddesdon
B379VWL	Dodge G13C	Saxon Sanbec	WrT	1984	Brill
Fire Alpha 2	Freezer Rigid	Boat & Trailer	FBt	1985	Marlow
C152DJO	Dodge G13	Saxon Sanbec	WrT	1985	Great Missenden
C153DJO	Dodge G13	Saxon Sanbec	WrT	1985	High Wycombe
D952RBW	Renault-Dodge G13		WrC	1987	Olney
D953RBW	Renault-Dodge S56	Rootes	CaV	1987	Aylesbury
D539XRX	Renault-Dodge G13	Saxon Sanbec	WrT	1987	Stokenchurch
E379DBL	Renault-Dodge G16C	Saxon Sanbec/Simon Snorkel SS70	HP	1987	Aylesbury
E914YWL	Renault-Dodge S66	Carmichael	CU	1987	Aylesbury
E447YWL	Volvo FL6.17	Mountain Range	WrT	1988	Bletchley
E448YWL	Volvo FL6.17	Mountain Range	WrT	1988	Chesham
F349FWL	GMC Scottsorle 3500 4x4	Mountain Range	BASU	1988	Waddesdon
F207GWL	GMC High Sierra 3500 4x4	Mountain Range	LWrT	1988	Broughton, Milton Keynes

One of the more unusual exhibits at the *Fire 95* exhibition in Harrogate was Buckinghamshire's K103UJO. Originally built as a Water Carrier for Buckinghamshire in 1993, it was rebuilt in 1995 by G B Fire to become the only Snozzle/Rescue Pump (SRP) in the UK. Modifications included a full crew cab, new body and the fitment of an American Snozzle hydraulic boom. At the head of the boom are a heat-seeking probe, floodlights and water monitor. *Gavin Stewart*

F312HWL	GMC High Sierra 3500 4x4	Telehoist	L4P	1988	Amersham
F66LBW	Leyland Freighter T45-180	Mountain Range	WrT	1989	Great Holm, Milton Keynes
F68LBW	Leyland Freighter T45-180	Mountain Range	RP	1989	Marlow
F311LBW	Bedford Brava 2300 4x4	Mountain Range/Bucks FRS	L4V	1989	Marlow
F312LBW	Bedford Brava 2300 4x4	Mountain Range/Bucks FRS	L4V	1989	Headquarters
F313LBW	Bedford Brava 2300 4x4	Mountain Range/Bucks FRS	L4V	1989	Headquarters
G511SBW	GMC 3500 4x4	Telehoist	LWrT	1989	Newport Pagnell
G458UBW	Dennis SS237	Mountain Range	RP	1990	Buckingham
G582YMJ	Land Rover 110	Land Rover	L4V	1989	Bletchley
G577YFC	Renault G170T	Chambers Engineering(1993) FoT/WrC		1990	Princess Risborough
H345FBW	Volvo FL6.17	Mountain Range	RP	1990	Bletchley
H812JUD	Leyland Freighter T45-180	Mountain Range	RP	1991	Aylesbury
H813JUD	Leyland Freighter T45-180	Mountain Range	RP	1991	Princess Risborough
H286UGP	Iveco Turbo-Daily 40.10 4x4	Chambers Engineering	HL	1991	Aylesbury
J53NFC	GMC 3500 4x4	Reliance Mercury	LWrT	1991	Beaconsfield
J857NJO	Scania G93ML250	Reliance Mercury	RP	1991	Broughton, Milton Keynes
J803TRO	Renault Midliner M230-15D	Emergency One	RP	1991	Beaconsfield
J712NWL	Volvo FL6.14	Chambers Engineering	OSU	1992	Aylesbury
J249OFC	Land Rover Defender 90 Tdi	Land Rover	L4V	1993	Aylesbury
J253OFC	Land Rover Defender 90 Tdi	Land Rover	L4V	1993	Headquarters
J376OWL	Land Rover Defender 90 Tdi	Land Rover	L4V	1993	High Wycombe
K102UJO	Volvo FL6.14	Chambers Engineering	OSU	1993	Bletchley
K103UJO	Volvo FL6.17	GB Fire/Snozzle P50(1995)	SRP	1993	Aylesbury
K104UJO	Volvo FL6.18	Emergency One	RP	1993	Amersham
K105UJO	Volvo FL6.18	Emergency One	RP	1993	Great Holm
K106UJO	Volvo FL7	Emergency One	RP	1993	High Wycombe
K107UJO	Volvo FL7	Emergency One	RP	1993	High Wycombe
K241VJO	Leyland-DAF 400 V8	Reliance Mercury/Bucks FRS	EPU	1993	Great Missenden
L129GMJ	Scania G93ML-210	Emergency One	RP	1993	Newport Pagnell
M387RWL	Renault Midliner M230-15D	Emergency One	RP	1994	Buckingham
M542MOG	Iveco-Ford 120E23	GB Fire/Magirus DLK24PLC	TL	1995	Newport Pagnall
M164SWL	Volvo FL6.18	Angloco/Simon Snorkel SS220	HP	1995	Beaconsfield
P982VUD	ERF EC8.24WT2	Saxon Sanbec/Emergency One	RP	1996	Gerrards Cross
P983VUD	ERF EC8.24WT2	Saxon Sanbec/Emergency One	RP	1996	
P984VUD	ERF EC8.24WT2	Saxon Sanbec/Emergency One	RP	1996	
P	Mercedes-Benz Unimog U2150		LWrT	1996	
P	Mercedes-Benz Unimog U2150		LWrT	1996	

Note: K103UJO and as rebuilt from a WrC in 1995. G577YFC received the body from K103UJO in 1995.

The first of the new ERF EC8 appliances were three Rescue Pumps for Buckinghamshire. These attractive units have crew cabs built by Saxon Sanbec and other appliance fittings by Emergency One. Shown at Gerrards Cross fire station is P982VUD.
Gavin Stewart

CAMBRIDGESHIRE FIRE & RESCUE SERVICE

Cambridgeshire Fire and Rescue Service, Hinchingbrook Cottage, Brampton Road, Huntingdon PE18 8NA

A727MEG	Dennis RS131	Dennis	WrL	1983	A24 Ramsey (R)
A728MEG	Dennis RS131	Dennis	WrT	1983	A24 Ramsey (R)
A729MEG	Dennis RS135	Dennis	WrL	1983	A28 St.Ives (R)
A621SEW	Iveco-Magirus 192D13	Carmichael / Magirus DL30	TL	1984	B01 Cambridge
A308VEG	Iveco-Magirus 192D13	Carmichael / Magirus DL30	TL	1984	A14 Dogsthorpe, Peterborough
B414JJE	Dennis RS135	Dennis	WrL	1984	Fire Provention
B415JJE	Dennis RS135	Dennis	WrT	1984	A22 Manea
B416JJE	Dennis RS135	Dennis	WrT	1984	A25 Sawtry
B56EFL	Dennis RS135	Dennis	WrL	1984	A17 Yaxley
B57EFL	Dennis RS135	Dennis	WrL	1984	A26 Kimbolton
B59EFL	Dennis RS135	Dennis	WrL	1984	A23 Chatteris
C103MEW	Dennis RS135	Dennis	WrL	1985	B07 Burwell
D146KVA	Dennis RS135	John Dennis	WrT	1986	A19 Thorney
D746KVA	Dennis RS134	HCB-Angus	WrL	1986	B06 Soham
D747KVA	Dennis RS135	HCB-Angus	WrT	1986	B04 Littleport
D748KVA	Dennis RS135	HCB-Angus	WrL	1986	A21 March
D749KVA	Dennis RS135	HCB-Angus	WrL	1986	A21 March
D750KVA	Bedford M1120 4x4	John Dennis	WrC	1986	A17 Yaxley
E739KEG	Dennis RS135	Fulton & Wylie	WrT	1987	B09 Linton
E740KEG	Dennis RS135	Fulton & Wylie	WrT	1987	Training Centre
E741KEG	Dennis RS135	Fulton & Wylie	WrL	1987	A18 Whittlesey
E742KEG	Dennis RS135	Fulton & Wylie	WrL	1987	A15 Peterborough Volunteers
E743KEG	Dennis RS135	Fulton & Wylie	WrT	1987	A15 Peterborough Volunteers
F808AAV	Dennis RS237	John Dennis	WrL	1988	Reserve
F809AAV	Dennis RS237	John Dennis	WrL	1988	Reserve
F810AAV	Dennis RS237	John Dennis	WrL	1988	Reserve
F811AAV	Dennis RS237	John Dennis	WrL	1988	Reserve
F611BAV	Volvo FL6.17	Fulton & Wylie	PM	1989	A27 Huntingdon
F612BAV	Volvo FL6.17	John Dennis (1992)/HIAB	HSRU	1989	B13 St.Neots
Pod 1	Fulton & Wylie	Incident Command & Control	ICCU	1989	A27 Huntingdon
G592OFL	Dennis RS237	John Dennis	WrL	1989	A28 St Ives
G593OFL	Dennis RS237	John Dennis	WrL	1989	B03 Sutton
G594OFL	Dennis RS237	John Dennis	WrL	1989	A24 Ramsey
G595OFL	Dennis RS237	John Dennis	WrL	1989	B02 Cottenham
H410CEW	Mercedes-Benz 811D	Carmichael	RU	1990	A14 Dogsthorpe, Peterborough
H491DFL	Dennis RS237	John Dennis	WrL	1990	Reserve
H492DFL	Dennis RS237	John Dennis	WrL	1990	A27 Huntingdon
H493DFL	Dennis RS237	John Dennis	WrL	1990	B13 St.Neots
H151JFL	Mercedes-Benz 1120AF	Rosenbauer	SRU	1991	A27 Huntingdon
J359SEW	Land Rover Defender 110	Land Rover/Cambs FRS	CU	1991	A14 Dogsthorpe, Peterborough
J988TEG	Dennis RS237	John Dennis	WrL	1991	A27 Huntingdon
J989TEG	Dennis RS237	John Dennis	WrL	1991	A20 Wisbech
J990TEG	Dennis RS237	John Dennis	WrL	1991	B13 St Neots
K681BEW	Dennis RS237	John Dennis	WrL	1992	A28 St Ives
K682BEW	Dennis RS237	John Dennis	WrL	1992	B08 Swaffham Bulbeck
L723NAV	Dennis RS237	John Dennis	WrL	1994	B11 Gamlingay
L577PFL	Dennis RS237	John Dennis	WrL	1994	Training Centre
L578PFL	Dennis RS237	John Dennis	WrL	1994	B10 Sawston
L351PFL	Mercedes-Benz 1124AF	Rosenbauer/Angloco	RDCU	1994	B01 Cambridge
L452TFL	Mercedes-Benz 814L	Gladwin	OSU	1994	A27 Huntingdon
L140VCE	Mercedes-Benz 2531	Carmichael International	FoT/WrC	1994	A27 Huntingdon
M846GEG	Dennis RS241	John Dennis	WrL	1995	B06 Soham
M847GEG	Dennis RS241	John Dennis	WrL	1995	B12 Papworth Everard
M848GEG	Dennis RS241	John Dennis	WrL	1995	A27 Huntingdon

This view shows Cambridgeshires Rescue/Damage Control Unit L351PFL. A Mercedes-Benz 1124AF chassis sports bodywork by Rosenbauer and Angloco that encapsulates considerable specialist equipment. When the bodyside shutters are raised, the body panels beneath them fold down forming sturdy steps. *Keith Grimes*

The penultimate batch of Dennis RSs for Cambridgeshire includes L723NAV seen here at Wisbech. After this batch of John Dennis-bodied pumps came a trio of similarly bodied RS241s, and more recently, four Dennis Sabres. An interesting feature of several newest pumps is the front mounted winch, prominent in this view. *Edmund Gray*

N489SAV	Dennis Sabre TSD233	John Dennis	WrL	1996	B13 St Neots
N490SAV	Dennis Sabre TSD233	John Dennis	WrL	1996	B05 Ely
N491SAV	Dennis Sabre TSD233	John Dennis	WrL	1996	A20 Wisbech
N492SAV	Dennis Sabre TSD233	John Dennis	WrL	1996	A27 Huntingdon
N812WAV	Volvo FL6.14	Leicestershire Carriage	CCU	1996	A27 Huntingdon
P150GFL	Dennis Sabre TSD233	John Dennis	WrL	1996	A16 Stanground, Peterborough
P151GFL	Dennis Sabre TSD233	John Dennis	WrL	1996	A14 Dogsthorpe, Peterborough
P152GFL	Dennis Sabre TSD233	John Dennis	WrL	1996	B01 Cambridge
P153GFL	Dennis Sabre TSD233	John Dennis	WrL	1996	B01 Cambridge
Boat			FBt	19..	A20 Wisbach

Note: F612BAV was originally a Prime Mover, it had the new HSRU body fitted in 1992.

CENTRAL SCOTLAND FIRE BRIGADE

Central Scotland Fire Brigade, Maddiston, Falkirk FK2 0LG

XLS278T	Dodge G1313	Fulton & Wylie	FSU	1979	F1 Bo'ness
A618HMS	Dodge G13C	Fulton & Wylie	WrT	1983	F3 Falkirk
A261JMS	Dodge G75C	Fulton & Wylie	SSU	1983	F3 Falkirk
B411MLS	Dodge S46C	Scott	CU	1985	F1 Bo'ness
B982MLS	Dodge S46	Scott/Fulton & Wylie	CaV	1985	C1 Alloa
B917RLS	Dodge G13C	Fulton & Wylie	WrL	1985	Reserve
B918RLS	Dodge G13C	Fulton & Wylie	WrL	1984	F1 Bo'ness
C872XLS	Dodge G13C	Fulton & Wylie	WrL	1985	S1 Aberfoyle
C873XLS	Dodge G13C	Fulton & Wylie	WrL	1985	S7 Killin
D242FMS	Renault-Dodge G16C	Fulton & Wylie	WrC/FoT	1987	S6 Dunblane
E111OLS	Volvo FL6.14	Fulton & Wylie	WrL	1988	Training Centre
E79TLS	Volvo FL6.14	Fulton & Wylie	WrL	1988	Reserve
E80TLS	Volvo FL6.14	Fulton & Wylie	WrL	1988	S5 Doune
E81TLS	Volvo FL6.14	Fulton & Wylie	WrT	1988	C1 Alloa
E82TLS	Volvo FL6.12	Aitken	GPV	1988	A2 Headquarters
F828BMS	Land Rover 110	Land Rover County	PCV	1989	A2 Headquarters
F994CMS	Volvo FL6.14	Mountain Range	WrL	1989	S3 Bridge of Allan
F995CMS	Volvo FL6.14	Mountain Range	WrL	1989	F5 Slamannan
F996CMS	Volvo FL6.14	Mountain Range	WrL	1989	F4 Larbert
F896VMS	Ford Transit	Fulton & Wylie	L2P	1989	V1 Crianlarich
G401HLS	Volvo FL6.14	Fulton & Wylie	WrT	1990	F2 Denny
G402HLS	Volvo FL6.14	Fulton & Wylie	WrT	1990	F1 Bo'ness
G403HLS	Volvo FL6.14	Fulton & Wylie	WrT	1990	S2 Balfron
H669OLS	Mercedes-Benz 917AF	Mountain Range/HIAB	SSU	1991	S8 Stirling

The latest delivery for the recently re-named Central Scotland Fire Brigade is this impressive command unit, N478VMS. A Volvo FL6.14 chassis is utilised built at the nearby Volvo factory in Irvine. The coachwork was also built in Scotland, at Cuper, by Heggie. When staff training is completed the machine is expected to be allocated to Bo'ness. *Andrew Fenton*

The Central Region Fire Brigade was renamed in April 1996 to become Central Scotland Fire Brigade. A pair of Specialist Support Units can be found at Falkirk and Stirling Fire Stations. The latter machine is **H699OLS** and carries Mountain Range bodywork on a Mercedes-Benz 917AF chassis. It was photographed passing through the town centre at Stirling in March 1996. *Tony Wilson*

H701OLS	Volvo FL6.14	Mountain Range	WrL	1991	S4 Callander
H702OLS	Volvo FL6.14	Mountain Range	WrL	1991	F2 Denny
H703OLS	Volvo FL6.14	Mountain Range	WrL	1991	C3 Tillycoultry
J366VLS	Volvo FL6.14	Emergency One	WrL	1992	F4 Larbert
J367VLS	Volvo FL6.14	Emergency One	WrL	1992	Reserve
J368VLS	Volvo FL6.14	Emergency One	WrL	1992	Reserve
J369VLS	Volvo FL6.14	Emergency One	WrL	1992	S6 Dunblane
J370VLS	Mercedes-Benz 711D	Fulton & Wylie	CP	1992	V2 Tyndrum
K456EMS	Volvo FL10	Angloco/Bronto Skylift 28.2Ti	ALP	1993	S8 Stirling
L801JLS	Volvo FL6.14	Emergency One	WrL	1993	F1 Bo'ness
L802JLS	Volvo FL6.14	Emergency One	WrL	1993	S8 Stirling
L803JLS	Volvo FL6.14	Emergency One	WrL	1993	F3 Falkirk
L804JVS	Mercedes-Benz 811D	Keillor	OHU	1993	A2 Headquarters
M858OLS	Volvo FL6.14	Emergency One	WrL	1994	C1 Alloa
N478VMS	Volvo FL6.14	Heggies	CU	1996	F1 Bo'ness
N480VMS	Volvo FL6.14	Emergency One	WrL	1996	F3 Falkirk
N481VMS	Volvo FL6.14	Emergency One	WrL	1996	S8 Stirling

CHESHIRE FIRE BRIGADE

Cheshire Fire Brigade, Winsford Compex, Sadler Road, Winsford, Cheshire CW7 2BN

Reg	Chassis	Body	Type	Year	Location
	Shand Mason Steamer	Shand Mason	P	1890	Preserved at Chester
LFM200	Dennis F7	Dennis	PE	1949	Preserved at Chester
RMB996	Dennis F8	Dennis	P	1953	Preserved at Macclesfield
NMB39P	Land Rover Series III 109FC	Carmichael Redwing	L4T	1975	12 Nantwich
EFM561S	Dennis Delta 2	Dennis	FoT	1978	08 Ellesmere Port
EFM562S	Dennis Delta 2	Dennis	FoT	1978	01 Warrington
WFM464W	Dennis F125	Dennis/Simon Snorkel SS263	HP	1980	Reserve
WFM465W	Dennis F125	Dennis/Simon Snorkel SS263	HP	1980	19 Macclesfield
DMB901X	Land Rover Series III 109 V8	Land Rover/Jennings	L4T	1981	13 Audlem
DMB902X	Land Rover Series III 109 V8	Land Rover/Jennings	L4T	1981	16 Sandbach
DMB903X	Land Rover Series III 109 V8	Land Rover/Jennings	CRU	1981	24 Knutsford
DMB904X	Land Rover Series III 109 V8	Land Rover/Jennings	L4T	1981	17 Holmes Chapel
DMB905X	Land Rover Series III 109 V8	Land Rover/Jennings	L4T	1981	10 Tarporley
DMB906X	Land Rover Series III 109 V8	Land Rover/Jennings	L4T	1981	22 Poynton
DMB907X	Land Rover Series III 109 V8	Land Rover/Jennings	L4T	1981	20 Bollington
DMB908X	Land Rover Series III 109 V8	Land Rover/Jennings	L4T	1981	11 Malpas
HFM710X	Dennis RS133	Dennis	HRU	1981	01 Warrington
A320RFM	Bedford CF2 350	Cheshire FB	SIU	1983	01 Warrington
A358VFM	Dennis RS133	Dennis	WrL	1984	Reserve
A359VFM	Dennis RS133	Dennis	WrL	1984	Reserve
A360VFM	Dennis RS133	Dennis	WrL	1984	08 Ellesmere Port
A361VFM	Dennis RS133	Dennis	WrL	1984	Training Centre
B272BMB	Dennis DF133	Carmichael/Magirus DL30	TL	1984	01 Warrington
B494CMB	Dennis DS151	Dennis	WrL	1985	Reserve
C100JCA	Dennis DS153	Dennis	WrL	1985	03 Stockton Heath
C101JCA	Dennis DS153	Dennis	WrL	1985	03 Stockton Heath
D876EVT	Bedford CF2	Bedford	GPV	1986	27 Winsford
D687PMB	Dennis SS133	Carmichael	WrL	1986	Reserve
D688PMB	Dennis SS133	Carmichael	WrL	1986	23 Wilmslow
D689PMB	Dennis SS133	Carmichael	WrL	1986	25 Middlewich
D690PMB	Dennis SS133	Carmichael	WrL	1986	Reserve
D691PMB	Dennis SS133	Carmichael	WrL	1986	13 Audlem
D692PMB	Dennis SS133	Carmichael	WrL	1986	12 Nantwich
E717SON	Scania P92M	Carmichael/Bronto Skylift 28-2Ti	APL	1987	09 Chester
E457XLG	Dennis SS135	Excalibur CBK	WrL	1987	11 Malpas
E458XLG	Dennis SS135	Excalibur CBK	WrL	1987	Reserve
E459XLG	Dennis SS135	Excalibur CBK	WrL	1987	22 Poynton
E460XLG	Dennis SS135	Excalibur CBK	WrL	1987	18 Congleton
E461XLG	Dennis SS135	Excalibur CBK	WrL	1987	Training school
E462XLG	Dennis SS135	Excalibur CBK	WrL	1987	12 Northwich
Trailer	SEB international	General Purpose	GPV	1988	09 Chester
F368KTU	GMC Custon Deluxe KC30 4x4	Macclesfield Motor Bodies	OSU	1989	15 Crewe
F369KTU	GMC Custon Deluxe KC30 4x4	Macclesfield Motor Bodies	OSU	1989	08 Ellesmere Port
Boat 4145	Chinook 3.8 Rigid Inflatable	Boat & Trailer	IRBt	1989	01 Warrington
Boat 4146	Chinook 3.8 Rigid Inflatable	Boat & Trailer	IRBt	1989	09 Chester
G998RFM	Leyland-DAF Roadrunner 8-15	Tractor unit	TU	1990	27 Winsford
Trailer	Gordons	Control Unit/Canteen Van	CU/CaV	1990	27 Winsford
G999RFM	Leyland-DAF Roadrunner 8-15	Tractor unit	TU	1990	25 Northwich
Trailer	Gordons	Control Unit/Canteen Van	CU/CaV	1990	25 Northwich
Trailer	Ifor Williams	General Purpose	GPV	1990	24 Knutsford
H442AMA	GMC 3500 4x4	Jennings	OSU	1991	01 Warrington
H443AMA	GMC 3500 4x4	Jennings	OSU	1991	C9 Congleton
J165NVM	Peugeot-Talbot Express	Talbot/Cheshire FB	GPV	1991	09 Chester
J421PVR	Peugeot-Talbot Express	Talbot/Cheshire FB	GPV	1992	25 Northwich
J422PVR	Peugeot-Talbot Express	Talbot/Cheshire FB	GPV	1992	27 Winsford

Opposite, top: **The fire appliance range produced by Dennis has been completely revised in the last couple of years. The Rapier model received a re-styled cab section similar to that used on the new Sabre model. Consequently, production of the RS and SS models has now ceased. Here is shown Cheshire's N712GFM, a new Sabre model.** *Gavin Stewart*

Opposite, bottom: **Articulated fire appliances are rare in Britain. Cheshire operate a pair of Leyland-DAF Road Runner tractors which take the Command and Control unit semi-trailers to incidents. Shown here is G998RFM, a unit based at Winsford fire station.** *Gavin Stewart*

Twenty Mercedes-Benz 1124F pumps preceded the Dennis Rapier WrLs in the Cheshire Fire Brigade. These arrived in two batches and were bodied by Locomotors and Carmichael International respectively. L969TFM is from the latter batch of six appliances. *Gavin Stewart*

J299HCA	Leyland-DAF 45-150	Lynton Commercials	SIU	1992	15 Crewe
J781HCA	Leyland-DAF 45-150	Lynton Commercials	SIU	1992	08 Ellesmere Port
K620KMB	Mercedes-Benz 1124F	Locomotors	WrL	1992	09 Chester
K621KMB	Mercedes-Benz 1124F	Locomotors	WrL	1992	01 Warrington
K622KMB	Mercedes-Benz 1124F	Locomotors	WrL	1992	08 Ellesmere Port
K623KMB	Mercedes-Benz 1124F	Locomotors	WrL	1992	06 Frodsham
K624KMB	Mercedes-Benz 1124F	Locomotors	WrL	1992	18 Congleton
K625KMB	Mercedes-Benz 1124F	Locomotors	WrL	1992	23 Wilmslow
K626KMB	Mercedes-Benz 1124F	Locomotors	WrL	1992	27 Winsford
K627KMB	Mercedes-Benz 1124F	Locomotors	WrL	1992	10 Tarporley
K628KMB	Mercedes-Benz 1124F	Locomotors	WrL	1992	16 Sandbach
K629KMB	Mercedes-Benz 1124F	Locomotors	WrL	1992	17 Holmes Chapel
K630KMB	Mercedes-Benz 1124F	Locomotors	WrL	1992	20 Bollington
K631KMB	Mercedes-Benz 1124F	Locomotors	WrL	1992	15 Crewe
K632KMB	Mercedes-Benz 1124F	Locomotors	WrL	1992	19 Macclesfield
K633KMB	Mercedes-Benz 1124F	Locomotors	WrL	1992	27 Winsford
L789VLE	Peugeot-Talbot Express	Talbot/Cheshire FB	GPV	1993	09 Chester
L968TFM	Mercedes-Benz 1124F	Carmichael International	WrL	1994	09 Chester
L969TFM	Mercedes-Benz 1124F	Carmichael International	WrL	1994	04 Widnes
L970TFM	Mercedes-Benz 1124F	Carmichael International	WrL	1994	19 Macclesfield
L971TFM	Mercedes-Benz 1124F	Carmichael International	WrL	1994	25 Northwich
L972TFM	Mercedes-Benz 1124F	Carmichael International	WrL	1994	05 Runcorn
L973TFM	Mercedes-Benz 1124F	Carmichael International	WrL	1994	15 Crewe
M58AMB	Ford Transit VE6	Ford	GPV	1995	01 Warrington
	Moffett Mounty All Terrain	*Normally carried by HFM710X*	FLT	1996	01 Warrington
N707GFM	Dennis Sabre S411	John Dennis	WrL	1996	24 Knutsford
N708GFM	Dennis Sabre S411	John Dennis	WrL	1996	08 Ellesmere Port
N709GFM	Dennis Sabre S411	John Dennis	WrL	1996	04 Widnes
N710GFM	Dennis Sabre S411	John Dennis	WrL	1996	05 Runcorn
N711GFM	Dennis Sabre S411	John Dennis	WrL	1996	02 Birchwood
N712GFM	Dennis Sabre S411	John Dennis	WrL	1996	09 Chester
N713GFM	Dennis Sabre S411	John Dennis	WrL	1996	01 Warrington
P	ERF		FoT	1996	
P	ERF		FoT	1996	

Note: HFM710X was previously WrL

CLEVELAND FIRE BRIGADE

Cleveland Fire Brigade, Stockton Road, Hartlepool, Cleveland TS25 5TB

2	D892DDC	Dennis DFS133	Saxon Sanbec	FoT	1986	3 Grangetown
3	H49KVN	Dennis DF237	John Dennis	ET/CU	1990	9 Stranton, Hartlepool
4	M812TVN	Dennis SS233	John Dennis	WrT	1995	1 Middlesbrough
5	A999HAJ	Shelvoke & Drewry WY	Carmichael / Bronto 72ft	HP	1984	9 Stranton, Hartlepool
6	N748YAJ	Dennis Sabre TSD233	John Dennis	WrT	1995	1 Middlesbrough
7	Boat	South Ocean Services	Cleveland Endeavour	FBt	1977	6 Marine, Middlesbrough
8	Boat	Marshall Branson	Cleveland Innovation	FBt	1992	6 Marine, Middlesbrough
9	H550NAJ	Dennis SS237	John Dennis	WrL	1991	14 Skelton
10	L679FPY	Dennis SS239	John Dennis	WrT	1993	9 Stranton, Hartlepool
12	M813TVN	Dennis SS239	John Dennis	WrT	1995	2 Stockton
13	WPY480Y	Dennis SS133	Dennis	WrT	1982	Training Centre
14	RHN896X	Dennis DF133	Dennis	FoT	1982	7 Billingham
16	M919SVN	Dennis SS239	John Dennis	ET/DCU	1995	8 Coulby Newham
17	L680FPY	Dennis SS239	John Dennis	WrT	1993	9 Stanton, Hartlepool
18	L681FPY	Dennis SS239	John Dennis	WrT	1993	2 Stockton
19	E727KAJ	Dennis SS135	John Dennis	WrT	1987	13 Saltburn
20	J125TVN	Steyr-Daimler-Puch Pinzgauer	Steyr-Daimler-Puch	L6R	1991	7 Billingham
22	N746YAJ	Dennis Sabre TSD233	John Dennis	WrT	1995	5 Thornaby
23	DDC1	Scania G93M-280	F&W/Simon ST300-S	HP	1991	2 Stockton
24	WPY482Y	Dennis SS133	Dennis	WrL	1982	Reserve
25	N750YAJ	Dennis Sabre TSD233	John Dennis	WrT	1995	9 Stanton, Hartlepool
26	M225PHN	Dennis DFS243	John Dennis/Atlas AS80.1	HRV	1995	2 Stockton
27	H551NAJ	Dennis SS237	John Dennis	WrT	1991	6 Marine, Middlesbrough
28	J126TVN	Steyr-Daimler-Puch Pinzgauer	Steyr-Daimler-Puch	L6RT	1991	3 Grangetown
29	C937ADC	Dennis SS135	John Dennis	WrT	1986	15 Loftus
30	C936ADC	Dennis SS135	John Dennis	WrL	1986	4 Redcar

Production of the Dennis RS/SS ranges ceased in 1995 in favour of the Rapier and Sabre models. Showing the distinctive white front unique to Cleveland, this John Dennis-bodied appliance runs from Coulby Newham. Numbered 16 (M919SVN) it was the last SS-type chassis built as a special appliance in this case a combined Emergency Tender and Damage Control Unit. *Karl Sillitoe*

Private index plates are less-common on fire appliances though Cleveland County Fire Brigade's 23 carries the mark DDC1. This was transferred from a Dennis appliance when the Scania was delivered. Simon Snorkel ST300-S booms were constructed upon the Scania G93M chassis, with Fulton & Wylie adding the bodywork. It is now on the run at Stockton Fire Station. *Karl Sillitoe*

31	E724KAJ	Dennis SS135	John Dennis	WrL	1987	12 Guisborough
32	E725KAJ	Dennis SS135	John Dennis	WrL	1987	7 Billingham
33	E726KAJ	Dennis SS135	John Dennis	WrL	1987	10 Headland, Hartlepool
34	J689VEF	Dennis SS237	John Dennis	WrT	1992	5 Thornaby
35	E604LHN	Scania P92M	Saxon Sanbec/Simon SS263	HP	1987	1 Middlesbrough
36	H552NAJ	Dennis SS237	John Dennis	WrL	1991	9 Stranton, Hartlepool
37	A988FAJ	Dennis SS133	Dennis	WrT	1983	Reserve
38	J688VEF	Dennis SS237	John Dennis	WrL	1992	8 Coulby Newham
39	M814TVN	Dennis SS239	John Dennis	WrT	1995	3 Grangetown
41	A989FAJ	Dennis SS133	Dennis	WrT	1983	Training Centre
42	B141RDC	Dennis SS135	Dennis	WrT	1985	Reserve
43	B142RDC	Dennis SS135	Dennis	WrL	1985	Reserve
45	B145LVN	Dennis SS133	Dennis	WrT	1984	11 Yarn
46	M226PHN	Dennis SS243	John Dennis	WrT	1994	1 Middlesbrough
47	M227PHN	Dennis SS243	John Dennis	WrT	1994	3 Grangetown
50	M985NVN	Renault Midliner M230	Renault	GPV	1994	1 Middlesbrough
51	J334THN	Land Rover Defender	Land Rover	L4V	1991	8 Coulby Newham
83	M264HPY	Ford Transit VE6	Ford	PCV	1994	Training Centre
85	J664SDC	Ford Transit VE6	Ford	PCV	1992	5 Thornaby
88	J332THN	Land Rover Defender	Land Rover	L4V	1991	4 Redcar
108	Caravan		Fire Prevention Unit	FPU	1986	Headquarters
141	Trailer	Whale Tankers	Foam Bowser	FoT	1977	7 Billingham
142	Trailer	Tow-a-Van			1977	7 Billingham
144	Boat	Searider	Rescue boat & trailer	IRBt	19	7 Billingham
145	Trailer	Merryweather Major	Monitor Trailer	Mon	1971	3 Grangetown
146	Trailer	Angus Armourlite	Oscillating Monitor Trailer	Mon	1983	7 Billingham

The Fire Brigade Handbook

CORNWALL COUNTY FIRE BRIGADE

Cornwall County Fire Brigade, County Hall, Station Road, Truro, TR1 3HA

A301	SGL770Y	Bedford TK	HCB-Angus	WrL	1983	5.2 St Columb
A302	A416WGL	Bedford TK	HCB-Angus	WrT	1983	5.1 Newquay
A303	A359YAF	Dodge G13C	HCB-Angus	WrL	1984	4.2 St Mawes
A304	B783BCV	Dodge G13C	HCB-Angus	WrL	1984	4.3 Mevagissey
A305	B784BCV	Dodge G13C	HCB-Angus	WrT	1984	3.1 Falmouth
A306	C107GAF	Dodge G13	HCB-Angus	WrL	1985	Reserve
A307	C109GAF	Dodge G13	HCB-Angus	WrL	1985	2.2 Redruth
A308	C110GAF	Dodge G13	HCB-Angus	WrL	1985	1.2 St Just
A309	B785BCF	Dodge G13C	HCB-Angus	WrL	1984	5.3 Padstow
A310	C108GAF	Dodge G13	HCB-Angus	WrT	1985	6.4 Lostwithiel
A311	D315KCV	Renault-Dodge G13T	HCB-Angus	WrT	1986	Training Centre
A312	D316KCV	Renault-Dodge G13T	HCB-Angus	WrL	1987	5.4 Wadebridge
A313	D317KCV	Renault-Dodge G13T	HCB-Angus	WrT	1986	3.2 St Keverne
A314	E444OGL	Renault-Dodge G13T	Saxon Sanbec	WrT	1987	7.4 Launceston
A315	E445OGL	Renault-Dodge G13T	Saxon Sanbec	WrT	1987	7.3 Bude
A316	E446OGL	Renault-Dodge G13T	Saxon Sanbec	WrL	1987	1.3 St Ives
B319	G743CAF	Renault-Dodge G13T	HCB-Angus	WrL	1990	7.2 Delabole
B320	G744CAF	Renault-Dodge G13T	HCB-Angus	WrL	1990	8.2 Looe
B321	G745CAF	Renault-Dodge G13T	HCB-Angus	WrL	1990	2.3 Perranporth
A324	F404URL	Renault-Dodge G13T	HCB-Angus	WrT	1988	6.2 St Dennis
A325	F405URL	Renault-Dodge G13T	HCB-Angus	WrL	1988	8.5 Callington
A326	F406URL	Renault-Dodge G13T	HCB-Angus	WrL	1988	8.4 Saltash
B327	G279CRL	Mercedes-Benz 1222F	HCB-Angus	WrT	1990	8.1 Liskeard
B328	G281CRL	Mercedes-Benz 1222F	HCB-Angus	WrL	1990	2.1 Camborne
B329	G282CRL	Mercedes-Benz 1222F	HCB-Angus	WrT	1990	6.1 St Austell
B330	G283CRL	Mercedes-Benz 1222F	HCB-Angus	WrL	1990	4.1 Truro

The latest deliveries to Cornwall County Fire Brigade comprise Mercedes-Benz 1124AFs with Carmichael International bodywork. The vehicles also feature a new livery that includes orange, yellow and blue reflective stripes that replace the previous red and yellow cheques. *Gavin Stewart*

Cornwall County Fire Brigade, like most brigades, was a user of Bedford chassied appliances. A brief flirtation with Dodge, latterly Renault-Dodge, ceased with the last batch delivered in 1989 as production was ending. This final batch of G13s have HCB-Angus bodywork, a supplier also to have ceased. A326 (F406URL) is based at Bodmin. *Peter Wilson*

B331	J612MAF	Mercedes-Benz 1120AF	Carmichael	WrL	1992	8.1 Liskeard
B332	J613MAF	Mercedes-Benz 1120AF	Carmichael	WrL	1992	6.1 St Austell
B333	J614MAF	Mercedes-Benz 1120AF	Carmichael	WrL	1992	1.1 Penzance
B334	J615MAF	Mercedes-Benz 1120AF	Carmichael	WrL	1992	1.3 St Ives
B335	L213VRL	Mercedes-Benz 1120AF	HCB-Angus	WrL	1994	7.3 Bude
B336	L214VRL	Mercedes-Benz 1120AF	HCB-Angus	WrL	1994	7.4 Launceston
B337	M561BAF	Mercedes-Benz 1120AF	Carmichael International	WrT	1994	2.1 Cambourne
B338	M563BAF	Mercedes-Benz 1120AF	Carmichael International	WrT	1994	7.1 Bodmin
B339	M114CAF	Mercedes-Benz 1120AF	Carmichael International	WrT	1994	4.1 Truro
B340	M113CAF	Mercedes-Benz 1120AF	Carmichael International	WrT	1994	6.3 Fowey
B350	L212VRL	Mercedes-Benz 1124F	HCB-Angus	WrL	1994	8.3 Torpoint
B351	L211VRL	Mercedes-Benz 1124F	HCB-Angus	WrL	1994	3.1 Falmouth
B352	M562BAF	Mercedes-Benz 1124F	Carmichael International	WrL	1994	5.1 Newquay
B353	M564BAF	Mercedes-Benz 1124F	Carmichael International	WrL	1994	7.1 Bodmin
B354	M112CAF	Mercedes-Benz 1124F	Carmichael International	WrL	1994	1.1 Penzance
B355	M115CAF	Mercedes-Benz 1124F	Carmichael International	WrL	1994	3.4 Helston
A359	WAF247T	Bedford TK	HCB-Angus	WrL	1978	8.4 Saltash
A360	CCV735V	Bedford TK	HCB-Angus	WrL	1979	Reserve
A361	CCV736V	Bedford TK	HCB-Angus	WrL	1979	Reserve
A362	NGL847X	Bedford TK	Cheshire Fire Engineering	WrL	1982	Training Centre
A363	NGL849X	Bedford TK	Cheshire Fire Engineering	WrL	1982	Training Centre
A364	PRL373Y	Bedford TK	HCB-Angus	WrL	1982	Reserve
A365	PRL374Y	Bedford TK	HCB-Angus	WrL	1982	Reserve
A366	RRL425Y	Bedford TK	HCB-Angus	WrL	1982	3.3 Mullion
A400	D51NAF	Renault-Dodge G16	Cornwall FB	WrC	1987	7.1 Bodmin
A402	E628RRL	Renault-Dodge G16	Cornwall FB	WrC	1988	1.1 Penzance
A405	F328XAF	Renault-Dodge G16	Cornwall FB	WrC	1989	2.1 Camborne
A407	NGL848X	Bedford TK	Cheshire Fire Engineering	ET	1982	7.1 Bodmin
A408	A987XCV	Bedford TK	HCB-Angus	ET	1984	2.1 Camborne
A409	A988XCV	Bedford TK	HCB-Angus	ET	1984	Reserve
B411	H625FGL	Renault-Dodge S75 4x4	Cornwall FB	M/WrT	1990	6.5 Polruan
A416	J752KAF	Kawasaki Mule 454 4x4	Cornwall FB	ATV	1991	8.2 Looe
A420	RCV273M	Dodge K1050	Cornwall CFB/Boniface (1994)	HRV	1974	2.1 Camborne
A422	L944UAF	Mercedes-Benz 814D	Cornwall FB	CU	1993	4.1 Truro

The Fire Brigade Handbook

The Hydraulic Platform at Truro illustrated in the last edition has been replaced by a totally new Aerial Ladder Platform during 1995. M568DAF has the rare combination of Mercedes-Benz 2531 chassis with Bedwas bodywork and Simon Snorkel ST290-S booms. The ST290-S variant has a working height of 29-metres, a similar appliance is on order for Newquay fire station, though this will have the 23-metre ST230-S booms. *Neil Faithfull*

A501	SRL771Y	Land Rover Series III 109	Cornwall FB	L4T	1983	Reserve
A502	SRL772Y	Land Rover Series III 109	Cornwall FB	L4T	1983	5.2 St Columb
A504	SRL774Y	Land Rover Series III 109	Cornwall FB	L4T	1983	3.4 Helston
A505	B132CAF	Land Rover 110	Cornwall FB	L4T	1985	8.4 Saltash
A506	B133CAF	Land Rover 110	Cornwall FB	L4T	1985	2.2 Redruth
A507	B134CAF	Land Rover 110	Cornwall FB	L4T	1985	6.4 Lostwithiel
A508	B135CAF	Land Rover 110	Cornwall FB	L4T	1985	2.3 Perranporth
A509	D672KRL	Land Rover 110	Cornwall FB	L4T	1986	7.2 Debabole
A510	D673KRL	Land Rover 110	Cornwall FB	L4T	1986	7.3 Bude
A511	E626RRL	Land Rover 110	Cornwall FB	L4T	1988	7.1 Bodmin
A512	SRL776Y	Land Rover Series III 109	Cornwall FB	L4T	1983	6.2 St Dennis
A513	E627RRL	Land Rover 110	Cornwall FB	L4T	1988	3.1 Falmouth
A514	D50NAF	Land Rover 110	Cornwall FB	L4T	1987	8.5 Callington
B515	G497CAF	Land Rover 110	Cornwall FB	L4T	1990	8.2 Looe
A516	D49NAF	Land Rover 110	Cornwall FB	L4T	1987	Reserve
B518	G940AGL	Land Rover 110	Cornwall FB	L4T	1989	1.1 Penzance
A521	YCV943T	Land Rover Series III 109	Carmichael	L4P	1979	8.3 Torpoint
A522	YCV945T	Land Rover Series III 109	Carmichael	L4P	1979	4.2 St Mawes
A523	ARL386T	Land Rover Series III 109	Carmichael	L4P	1979	1.2 St Just
A524	YCV944T	Land Rover Series III 109	Carmichael	L4P	1979	5.4 Wadebridge
A525	YGL138T	Land Rover Series III 109	Carmichael	L4P	1979	1.3 St Ives
A526	ECV853V	Land Rover Series III 109	Carmichael	L4P	1980	3.2 St Keverne
A528	ECV855V	Land Rover Series III 109	Carmichael	L4P	1980	5.3 Padstow
A529	ECV856V	Land Rover Series III 109	Carmichael	L4P	1980	4.3 Mevagissey
A530	FAF92V	Land Rover Series III 109	Carmichael	L4P	1980	3.3 Mullion
A531	FAF93V	Land Rover Series III 109	Carmichael	L4P	1980	6.3 Fowey
B540	N376LAF	Vauxhall Brava 4x4	Vauxhall/CCFB	L4P	1996	Workshops
	M568DAF	Mercedes-Benz 2531	Bedwas/Simon Snorkel ST290-S	ALP	1995	4.1 Truro
	N94LCV	Mercedes-Benz 1124F	Carmichael International	WrT	1997	8.5 Calington
	N95LCV	Mercedes-Benz 1124AF	Carmichael International	WrT	1997	8.4 Saltash
	P136PCV	Mercedes-Benz 1124AF	Carmichael International	RT	1996	7.1 Bodmin
	P137PCV	Mercedes-Benz 1827	Simon/Simon Snorkel ST240-S	ALP	1997	5.1 Newquay
	P138PCV	Mercedes-Benz 1124AF	Carmichael International	RT	1996	2.1 Cambourne
	P156SCY	Mercedes-Benz 1124F	Carmichael International	WrT	1997	On order for St Mary, Isle of Scilly

CUMBRIA FIRE SERVICE

Cumbria Fire Service, Station Road, Cockermouth, Cumbria CA13 9PR

01-24	WHH29Y	Dodge G075	Carmichael	WrT	1983	Arnside
01-26	WHH31Y	Dodge G075	Carmichael	WrT	1983	Reserve
01-28	WHH33Y	Dodge G075	Carmichael	WrT	1983	Reserve
01-29	A752DRM	Dodge G12	Carmichael	WrT	1984	Kirkby Stephen
01-30	A753DRM	Dodge G12	Carmichael	WrT	1984	Reserve
01-31	A754DRM	Dodge G13	Carmichael	WrT	1984	Egremont
01-32	A755DRM	Dodge G12	Carmichael	WrT	1984	Sedburgh
01-33	A756DRM	Dodge G12	Carmichael	WrT	1984	Appleby
01-34	B22HRM	Dodge G13	Carmichael	WrT	1985	Wigton
01-35	B23HRM	Dodge G13	Carmichael	WrT	1985	Walney
01-36	B24HRM	Dodge G13	Carmichael	WrT	1985	Windermere
01-37	B25HRM	Dodge G13	Carmichael	WrT	1985	Keswick
01-38	B26HRM	Dodge G13	Carmichael	WrT	1985	Keswick
01-39	C853MAO	Dodge G13	Carmichael	WrT	1986	Reserve
01-40	C854MAO	Dodge G13	Carmichael	WrT	1986	Reserve
01-41	C855MAO	Dodge G13	Carmichael	WrT	1986	Longtown
01-42	C856MAO	Dodge G13	Carmichael	WrT	1986	Reserve
01-43	C857MAO	Dodge G13	Carmichael	WrT	1986	Seascale
01-44	D97UHH	Renault-Dodge G13	Carmichael	WrT	1987	Shap
01-45	D98UHH	Renault-Dodge G13	Carmichael	WrT	1987	Silloth
01-46	D99UHH	Renault-Dodge G13	Carmichael	WrL	1987	Windermere
01-47	E994AHH	Volvo FL6.14	Carmichael	WrT	1988	Brampton
01-48	E995AHH	Volvo FL6.14	Carmichael	WrT	1988	Ulverston
01-49	E996AHH	Volvo FL6.14	Carmichael	WrL	1988	Ulverston
01-50	F111HAO	Volvo FL6.14	Carmichael	WrT	1988	Dalton-in-Furness
01-51	F112HAO	Volvo FL6.14	Carmichael	WrT	1989	Staveley
01-52	F113HAO	Volvo FL6.14	Carmichael	WrT	1988	Maryport
01-53	F114HAO	Volvo FL6.14	Carmichael	WrT	1989	Maryport
01-54	G791OAO	Volvo FL6.14	Carmichael	WrT	1990	Grange over Sands
01-55	G792OAO	Volvo FL6.14	Carmichael	WrT	1990	Penrith
01-56	G793OAO	Volvo FL6.14	Carmichael	WrT	1990	Whitehaven
01-57	H481TRM	Volvo FL6.14	Carmichael	WrT	1991	Grange over Sands
01-58	H484TRM	Volvo FL6.14	Carmichael	WrT	1991	Kendal
01-59	H479TRM	Volvo FL6.14	Carmichael	WrL	1991	Penrith
01-60	H485TRM	Volvo FL6.14	Carmichael	WrL	1991	Workington
01-61	H483TRM	Volvo FL6.14	Carmichael	WrT	1991	Milnthorpe
01-62	H487TRM	Volvo FL6.14	Carmichael	WrT	1991	Whitehaven
01-63	J760XHH	Leyland-DAF FA45-150	Carmichael	C/WrL	1992	Coniston
01-64	K477FRM	Leyland-DAF FA45-160	HCB-Angus	C/WrL	1993	Lazonby
01-65	L214KAO	Leyland-DAF FA45-160	HCB-Angus	C/WrL	1994	Millom
01-66	L215KAO	Leyland-DAF FA45-160	HCB-Angus	C/WrL	1994	Broughton in Furness

The Cumbria fleet numbering system allocates a two number code to each specific type of appliance, followed by an individual number for each vehicle. Type 05 is for Rescue Units and Major Rescue Vehicles, and one of the latter is 05-07 (H486TRM) at Workington.
Malcolm Cook

01-67	L216KAO	Leyland-DAF FA45-160	HCB-Angus	C/WrL	1994	Bootle
01-68	L217KAO	Leyland-DAF FA45-160	HCB-Angus	C/WrL	1994	Millom
01-69	L218KAO	Leyland-DAF FA45-160	HCB-Angus	C/WrL	1994	Patterdale
01-70	L85MAO	Volvo FL6.11	Carmichael International	C/WrL	1994	Ambleside
01-71	L86MAO	Volvo FL6.11	Carmichael International	C/WrL	1994	Alston
01-72	L87MAO	Volvo FL6.11	Carmichael International	C/WrL	1994	Kirkby Lonsdale
01-73	L89MAO	Volvo FL6.11	Carmichael International	C/WrL	1994	Frizington
01-74	L91MAO	Volvo FL6.11	Carmichael International	C/WrL	1994	Aspatria
01-75	M246RRM	Volvo FL6.14	Emergency One	WrL	1995	Barrow-in-Furness
01-76	M247RRM	Volvo FL6.14	Emergency One	WrL	1995	Coackermouth
01-77	M248RRM	Volvo FL6.14	Emergency One	WrL	1995	Barrow-in-Furness
01-78	M249RRM	Volvo FL6.14	Emergency One	WrL	1995	Carlisle
01-79	N527WRM	Volvo FL6.14	Emergency One	WrL	1996	Carlisle
01-80	N528WRM	Volvo FL6.14	Emergency One	WrL	1996	Workington
01-81	N529WRM	Volvo FL6.14	Emergency One	WrL	1996	Keswick
01-82	P	Volvo FL6.14	Emergency One	WrL	1996/97	
01-83	P	Volvo FL6.14	Emergency One	WrL	1996/97	
01-84	P	Volvo FL6.14	Emergency One	WrL	1996/97	
01-85	P	Volvo FL6.14	Emergency One	WrL	1996/97	
03-01	E998AHH	Volvo FL6.16	Carmichael/Magirus DL30	TL	1988	Carlisle
03-02	E999AHH	Volvo FL6.16	Carmichael/Magirus DL30	TL	1988	Barrow-in-Furness
03-03	G790OAO	Volvo FL6.17	Carmichael/Magirus DL30	TL	1990	Workington
05-03	A92EAO	Dodge G12	Ecrolite	RU	1984	Barrow in Furness
05-04	B652HRM	Dodge G12C	Ecrolite	RU	1985	Reseve
05-05	B713KRM	Dodge G12	Ecrolite	RU	1985	Penrith
05-06	D426TAO	Renault-Dodge G12	Ecrolite	RU	1987	Kendal
05-07	H486TRM	Volvo FL6.14	Carmichael	MRV	1991	Workington
05-08	N530WRM	Volvo FL6.14	Emergency One	MRV	1996	Carlisle
05-09	P	Volvo FL6.14	Emergency One	MRV	1996/97	
06-01	D539TRM	Renault-Dodge G11	Robin Hood	ISU	1987	Kendal
06-02	D540TRM	Renault-Dodge G11	Robin Hood	ISU	1987	Whitehaven
07-02	E993AHH	Volvo FL6.14	Carmichael	FoT	1988	Carlisle
07-03	E997AHH	Volvo FL6.14	Carmichael	FoT	1988	Whitehaven
07-04	H482TRM	Volvo FL6.14	Carmichael	FoT	1991	Barrow-in-Furness
09-01	B653HRM	Land Rover 110	Carmichael	L4P	1985	Sedburgh
09-02	C591MHH	Land Rover 110	Carmichael	L4P	1986	Ambleside
09-03	C877PRM	Land Rover 110	Carmichael	L4P	1986	Penrith
09-04	C878PRM	Land Rover 110	Carmichael	L4P	1986	Appleby
13-01	Pod	Penman	Control unit	CU	1987	Headquarters
24-01	D93UHH	Renault-Dodge G11	Ray Smith	PM	1987	Workington
24-02	D94UHH	Renault-Dodge G11	Ray Smith	PM	1987	Headquarters
24-03	D95UHH	Renault-Dodge G11	Ray Smith	PM	1987	Carlisle
24-04	D96UHH	Renault-Dodge G11	Ray Smith	PM	1987	Barrow-in-Furness
0	Pod	Penman	BA training unit	BA(t)	1987	Training Centre
0	Pod	Penman	BA training unit	BA(t)	1987	Training Centre
0	Pod	Ray Smith	Flat bed lorry	FBL	1987	Carlisle
0	Pod	Ray Smith	Flat bed lorry	FBL	1987	Reserve
0	Pod	Ray Smith	Flat bed lorry	FBL	1987	Workington

Cumbria's latest appliances have coachwork by Emergency One on the popular Volvo F6 chassis, now a standard choice for several brigades. 01-79 (N529WRM) is a Water-Tender Ladder from the 1996 delivery, it is based at Carlisle.
Robert Smith

Derbyshire Fire Service took delivery of two new Emergency Tenders in 1995. The appliances went on the run at Buxton and Ripley. The Buxton, M551TET, machine is illustrated here. Both machines have Angloco bodywork allied to Renault Midliner chassis. *Clive Shearman*

Having purchased an initial pair of the original style Dennis Rapiers delivered in 1992, Derbyshire took ten of the revised model in 1995. Allocated throughout the county, these have Carmichael International bodies and were originally fitted with labour saving ladder gantries now removed. Based at Buxton, M547TET illustrates the type in original condition. *Clive Shearman*

DERBYSHIRE FIRE SERVICE

Derbyshire Fire Service, Old Hall, Burton Road, Littleover, Derby DE3 6EH

BA05	C45ORA	Dennis DF135	Cottingham/Campair	BAT	1986	Stored
CU02	D999URC	Dodge G13	Customline	CU	1986	Ripley
CV02	G999PTV	Leyland-DAF Roadrunner	Frank Guy	CaV	1990	Clay Cross
DC01	E998DCH	Dennis SS135	Cottingham	DecU	1988	Chesterfield
DC02	E999DCH	Dennis SS135	Cottingham	DecU	1988	Ascot Drive, Derby
ET01	M551TET	Renault Midlander M230-15D	Angloco	ET	1984	Buxton
ET02	M552TET	Renault Midlander M230-15D	Angloco	ET	1984	Ripley
EX01	Trailer		Exhibition Unit	ExU	1996	Fire Safety
EX02	Trailer		Exhibition Unit	ExU	1996	Fire Safety
HP03	G999NCH	Dennis F127	Saxon / Simon Snorkel ST220-S	HP	1990	Ascot Drive, Derby
HP04	G999NRB	Dennis F127	Saxon / Simon Snorkel ST240-S	HP	1990	Buxton
L4P4	H682SWG	Mercedes-Benz Unimog U1250L	Carmichael	L4P	1990	Matlock
LC63	L717NKU	Leyland-DAF 50	Leyland	DTV	1994	Driving School
LC64	F999HRB	Leyland-Freighter T45	Leyland	DTV	1988	Driving School
LR14	H593TDT	Land Rover Defender 90	Land Rover	L4V	1991	Wirksworth
LR15	H594TDT	Land Rover Defender 90	Land Rover	L4V	1991	Bakewell
LR23	N802BDT	Land Rover Defender 110	Land Rover	CU	1995	Buxton
LR24	N794BDT	Land Rover Defender 110	Land Rover	LiRU	1995	Matlock
LR86	K806GWG	Land Rover Defender 90	Land Rover	L4V	1993	Crich
LR87	K807GWG	Land Rover Defender 90	Land Rover	L4V	1993	Bolsover
LR88	K808GWG	Land Rover Defender 90	Land Rover	L4V	1993	Fire Safety
LR91	L891MWG	Land Rover Defender 90	Land Rover	L4V	1994	Whaley Bridge
LR96	J696CWB	Land Rover Defender 90	Land Rover	L4V	1992	Glossop
LR97	J697CWB	Land Rover Defender 90	Land Rover	L4V	1992	Bradwell
LR98	J698CWB	Land Rover Defender 90	Land Rover	L4V	1992	Hathersage
LR99	J695CWB	Land Rover Defender 90	Land Rover	L4V	1992	Clowne
MB57	L907MWG	Leyland DAF 200	Leyland DAF	PCV	1994	Buxton
MB58	L908MWG	Leyland DAF 200	Leyland DAF	PCV	1994	Kingsway, Derby
MB59	L903MWG	Leyland DAF 200	Leyland DAF	PCV	1994	Ascot Drive, Derby
MB60	L890MWG	Leyland DAF 400	Leyland DAF	PCV	1994	Training Centre
MB61	L901MWG	Leyland DAF 400	Leyland DAF	PCV	1994	Training Centre
MB64	L714NKU	Leyland DAF 200	Leyland DAF	PCV	1994	Chesterfield
MB66	M466VWE	LDV 200	LDV	PCV	1995	Duffield
TU07	B999KTV	Shelvoke & Drewry WY	Carmichael / Magirus DL30	TL	1985	Reserve
TU08	B998KTV	Shelvoke & Drewry WY	Carmichael / Magirus DL30	TL	1985	Kingsway, Derby
TU09	E999BRB	Renault-Dodge G16	Carmichael / Magirus DL30	TL	1988	Chesterfield
WC03	OVO83W	Dodge G1613	Angloco / Butterfield	WrC	1982	Ilkeston
WC04	ACH449Y	Dennis DF133	Dennis Waterbird	WrC	1983	Chesterfield
WC05	E999BNN	Bedford TM 4x4	Carmichael	WrC	1987	Buxton
WL01	L891MWB	Dennis SS237	Carmichael	WrL	1993	Alfreton
WL02	L892MWB	Dennis SS237	Carmichael	WrL	1993	Chesterfield
WL03	L893MWB	Dennis SS237	Carmichael	WrL	1993	Nottingham Rd, Derby
WL04	M544TET	Dennis Rapier TF203	Carmichael International	WrL	1995	Ilkeston
WL05	M545TET	Dennis Rapier TF203	Carmichael International	WrL	1995	Long Eaton
WL06	M546TET	Dennis Rapier TF203	Carmichael International	WrL	1995	Ripley
WL07	M547TET	Dennis Rapier TF203	Carmichael International	WrL	1995	Buxton
WL08	M548TET	Dennis Rapier TF203	Carmichael International	WrL	1995	Glossop
WL09	M469VWE	Dennis Rapier TF203	Carmichael International	WrL	1995	New Mills
WL10	M445VWE	Dennis Rapier TF203	Carmichael International	WrL	1995	Matlock
WL11	M461VWE	Dennis Rapier TF203	Carmichael International	WrL	1995	Ascot Drive, Derby
WL12	M462VWE	Dennis Rapier TF203	Carmichael International	WrL	1995	Kingsway, Derby
WL13	M463VWE	Dennis Rapier TF203	Carmichael International	WrL	1995	Swadlincote
WL55	VRC996Y	Dennis RS131	Dennis	WrL	1982	Training Centre
WL56	VRC997Y	Dennis RS133	Dennis	WrL	1982	Training Centre
WL57	VRC998Y	Dennis RS133	Dennis	WrL	1982	Training Centre
WL58	VRC999Y	Dennis RS133	Dennis	WrL	1982	Shirebrook
WL59	A994CCH	Dennis RS133	Dennis	WrL	1983	Melbourne
WL60	A995CCH	Dennis RS133	Dennis	WrL	1983	Chapel-en-le-Frith
WL61	A996CCH	Dennis RS133	Dennis	WrL	1983	Buxton
WL62	A997CCH	Dennis RS133	Dennis	WrL	1983	Ashbourne
WL63	A998CCH	Dennis RS133	Dennis	WrL	1983	Ripley
WL64	A999CCH	Dennis RS133	Dennis	WrL	1983	Swadlincote
WL65	D444TRB	Dennis RS135	Carmichael	WrL	1986	Bakewell
WL66	D445TRB	Dennis RS135	Carmichael	WrL	1986	Bolsover

WL67	D446TRB	Dennis RS135	Carmichael	WrL	1986	Reserve
WL68	D447TRB	Dennis RS135	Carmichael	WrL	1986	Reserve
WL69	D448TRB	Dennis RS135	Carmichael	WrL	1986	New Mills
WL70	D449TRB	Dennis RS135	Carmichael	WrL	1986	Hathersage
WL71	D450TRB	Dennis RS135	Carmichael	WrL	1986	Clay Cross
WL72	D991XAU	Dennis SS135	Carmichael	WrL	1987	Duffield
WL73	D992XAU	Dennis SS135	Carmichael	WrL	1987	Wirksworth
WL74	D993XAU	Dennis SS135	Carmichael	WrL	1987	Matlock
WL75	D994XAU	Dennis SS135	Carmichael	WrL	1987	Crich
WL76	D995XAU	Dennis SS135	Carmichael	WrL	1987	Ilkeston
WL77	D996XAU	Dennis SS135	Carmichael	WrL	1987	Chesterfield
WL78	D997XAU	Dennis SS135	Carmichael	WrL	1987	Heanor
WL79	D998XAU	Dennis SS135	Carmichael	WrL	1987	Glossop
WL80	D999XAU	Dennis SS135	Carmichael	WrL	1987	Reserve
WL81	D999XCH	Dennis SS135	Carmichael	WrL	1987	Long Eaton
WL82	G992NCH	Dennis SS135	Carmichael	WrL	1989	Staveley
WL83	G993NCH	Dennis SS135	Carmichael	WrL	1989	Whaley Bridge
WL84	G994NCH	Dennis SS135	Carmichael	WrL	1989	Reserve
WL85	G995NCH	Dennis SS135	Carmichael	WrL	1989	Reserve
WL86	G996NCH	Dennis SS135	Carmichael	WrL	1989	Reserve
WL87	G997NCH	Dennis SS235	Carmichael	WrL	1989	Alfreton
WL88	G998NCH	Dennis SS235	Carmichael	WrL	1989	Reserve
WL89	J558XHL	Dennis SS237	Carmichael	Wrl	1991	Driving School
WL90	J559XHL	Dennis SS237	Carmichael	WrL	1991	Bradwell
WL91	J561XHL	Dennis SS237	Carmichael	WrL	1991	Ashbourne
WL92	J562XHL	Dennis SS237	Carmichael	WrL	1991	Training Centre
WL93	J563XHL	Dennis SS237	Carmichael	WrL	1991	Ascot Drive, Derby
WL94	J564XHL	Dennis SS237	Carmichael	WrL	1991	Shirebrook
WL95	J565XHL	Dennis SS237	Carmichael	WrL	1991	Chesterfield
WL96	K936EWG	Dennis Rapier TF203	Carmichael	WrL	1992	Dronfield
WL97	K937EWG	Dennis Rapier TF203	Carmichael	WrL	1992	Clowne
WL98	K938EWG	Dennis SS237	Carmichael	WrL	1992	Clay Cross
WL99	K935EWG	Dennis SS237	Carmichael	WrL	1992	Belper

Opposite, top: **One of the smaller Aerial Ladder Platform appliances is the 24-metre Simon ST240-S seen here on Derbyshire's G999NRB at Buxton. The chassis is a Dennis F127, a type not often supplied for an ALP.** *Clive Shearman*

DEVON FIRE & RESCUE SERVICE

Devon Fire & Rescue Service, Clyst St George, Exeter EX3 0NW

5010	CDV77Y	Ford A0609/NAM 4x4	Fulton & Wylie	HL	1982	33 Exmouth
5012	CTT284Y	Dennis RS133	Dennis	WrL	1982	13 Okehampton
5015	CTT287Y	Dennis RS133	Dennis	WrL	1982	Training Centre
5016	CTT288Y	Dennis RS133	Dennis	WrL	1982	31 Totnes
5017	CTT289Y	Dennis RS133	Dennis	WrL	1982	Reserve
5022	A661KFJ	Dennis RS133	Dennis	WrL	1983	Training Centre
5024	A662KFJ	Dennis RS133	Dennis	WrL	1983	Reserve
5025	A663KFJ	Dennis RS133	Dennis	WrL	1983	Reserve
5027	A664KFJ	Dennis RS133	Dennis	WrL	1983	45 Topsham
5031	A665KFJ	Dennis RS133	Dennis	WrL	1983	Reserve
5032	A666KFJ	Dennis RS133	Dennis	WrL	1983	40 Honiton
5035	A667KFJ	Dennis RS133	Dennis	WrL	1983	44 Tiverton
5047	A974NDV	Dodge RG12	Angloco	ET	1984	01 Barnstaple
5048	A975NDV	Dodge RG12	Angloco	ET	1984	17 Torquay
5049	B873TTA	Dennis RS135	Dennis	WrL	1984	Reserve
5050	B874TTA	Dennis RS135	Dennis	WrL	1984	Reserve
5051	B875TTA	Dennis RS135	Dennis	WrL	1984	18 Paignton
5052	B876TTA	Dennis RS135	Dennis	WrL	1984	Reserve
5053	B877TTA	Dennis RS135	Dennis	WrL	1984	Reserve
5054	B317VFJ	Dodge S75/Boughton 4x4	Angloco	HL	1984	17 Torquay
5055	B318VFJ	Iveco Magirus 256D14	Magirus DLK23.12	TL	1984	Reserve
5097	C710FOD	Dodge RG13	Saxon Sanbec/Devon FRS	HL	1986	49 Crownhill
5129	D616NDV	Dennis RS135	Saxon Sanbec	WrL	1986	57 Tavistock
5130	D617NDV	Dennis RS135	Saxon Sanbec	WrL	1986	40 Honiton

Opposite: **Eighteen MAN L2000 County Water Tender Ladders were supplied to Devon in 1995. The compact dimensions, as shown by M59KFJ, make the type ideal for the rural roads of Devon.** *R Smith*

5131	D618NDV	Dennis RS135	Saxon Sanbec	WrL	1986	34 Axminster
5132	D619NDV	Dennis RS135	Saxon Sanbec	WrL	1986	33 Exmouth
5133	D620NDV	Dennis RS135	Saxon Sanbec	WrL	1986	Driver Training
5134	D621NDV	Dennis RS135	Saxon Sanbec	WrL	1987	24 Dartmouth
5135	D622NDV	Dennis RS135	Saxon Sanbec	WrL	1987	10 Holsworthy
5140	E488VDV	Dennis RS135	Angloco	WrL	1988	13 Okehampton
5141	E489VDV	Dennis RS135	Angloco	WrL	1988	25 Dawlish
5142	E995VTA	Dennis RS135	Angloco	WrL	1988	31 Totnes
5143	E996VTA	Dennis RS135	Angloco	WrL	1988	02 Ilfracombe
5144	E997VTA	Dennis RS135	Angloco	WrL	1988	09 Hatherleigh
5145	E998VTA	Dennis RS135	Angloco	WrL	1988	44 Tiverton
5146	E999VTA	Dennis RS135	Angloco	WrL	1988	17 Torquay
5168	LTT197P	Commer Commando RG11	HCB-Angus/Devon FRS	ET	1975	Reserve
5189	B324XTT	Dennis RS135	Dennis	WrL	1985	38 Crediton
5190	B325XTT	Dennis RS135	Dennis	WrL	1985	Driver Training
5191	B326XTT	Dennis RS135	Dennis	WrL	1985	Reserve
5192	B327XTT	Dennis RS135	Dennis	WrL	1985	Driver Trainer
5193	B328XTT	Dennis RS135	Dennis	WrL	1985	43 Sidmouth
5194	B329XTT	Dennis RS135	Dennis	WrL	1985	Reserve
5195	B330XTT	Dennis RS135	Dennis	WrL	1985	Driver Trainer
5241	F193EDV	Dennis RS135	John Dennis	WrL	1988	05 Braunton
5242	F194EDV	Dennis RS135	John Dennis	WrL	1988	51 Plymstock
5243	VDR291V	Dodge RG11	Benson	ET	1979	49 Crowhill
5244	VDR292V	Dodge RG11	Benson	ET	1979	32 Exeter
5247	F195EDV	Dennis RS135	John Dennis	WrL	1988	32 Exeter
5248	F196EDV	Dennis RS135	John Dennis	WrL	1988	38 Crediton
5251	F197EDV	Dennis RS135	John Dennis	WrL	1988	43 Sidmouth
5252	F198EDV	Dennis RS135	John Dennis	WrL	1988	57 Tavistock
5253	F199EDV	Dennis RS135	John Dennis	WrL	1988	28 Newton Abbott
5260	WTK454V	Dodge G11	HCB-Angus	WrL	1980	Training Centre
5265	WTK459V	Dodge G11	HCB-Angus	WrL	1980	56 Princetown
5304	PFJ274W	Ford A0609/NAM 4x4	G&T Fire Control	HL	1981	01 Barnstaple
5308	F650JFJ	Dennis F127	John Dennis/Camiva EMA30	TL	1989	01 Barnstaple
5323	G76OTA	Volvo FL6.14	Saxon Sanbec	WrL	1989	46 Witheridge
5324	G77OTA	Volvo FL6.14	Saxon Sanbec	WrL	1989	32 Exeter
5325	G78OTA	Volvo FL6.14	Saxon Sanbec	WrL	1989	04 Bideford
5326	G79OTA	Volvo FL6.14	Saxon Sanbec	WrL	1989	04 Bideford
5327	G80OTA	Volvo FL6.14	Saxon Sanbec	WrL	1990	30 Teignmouth
5328	G81OTA	Volvo FL6.14	Saxon Sanbec	WrL	1990	30 Teignmouth
5329	G82OTA	Volvo FL6.14	Saxon Sanbec	WrL	1990	39 Cullompton
5350	H179AFJ	Dennis RS237	John Dennis	WrL	1990	17 Torquay
5351	H180AFJ	Dennis RS237	John Dennis	WrL	1990	48 Camels Head
5352	H181AFJ	Dennis RS237	John Dennis	WrL	1990	17 Torquay
5353	H182AFJ	Dennis RS237	John Dennis	WrL	1990	50 Greenbank
5354	H183AFJ	Dennis RS237	John Dennis	WrL	1990	18 Paignton
5355	H184AFJ	Dennis RS237	John Dennis	WrL	1990	28 Newton Abbot
5356	H185AFJ	Dennis RS237	John Dennis	WrL	1990	02 Ilfracombe
5357	H186AFJ	Dennis RS237	John Dennis	WrL	1990	33 Exmouth
5358	H187AFJ	Dennis RS237	John Dennis	WrL	1990	45 Topsham
5359	H188AFJ	Volvo FL10	Angloco/Bronto Skylift 28-2Ti	HP	1990	48 Camels Head
5366	J862MFJ	Dennis DS155	John Dennis	WrL	1991	Reserve
5367	J863MFJ	Dennis DS155	John Dennis	WrL	1991	20 Bovey Tracey
5368	J864MFJ	Dennis DS155	John Dennis	WrL	1991	24 Dartmouth
5369	J865MFJ	Dennis DS155	John Dennis	WrL	1991	Reserve
5370	J866MFJ	Dennis DS155	John Dennis	WrL	1991	27 Moretonhampstead
5371	J867MFJ	Dennis DS155	John Dennis	WrL	1991	58 Yelverton
5374	J868MFJ	Dennis DS155	John Dennis	WrL	1991	21 Brixham
5383	J869MFJ	Volvo FL10	Angloco / Bronto Skylift 28-2Ti	HP	1991	32 Exeter
5400	J688MTA	Volvo FL10	Angloco / Bronto Skylift 28-2Ti	HP	1992	17 Torquay
5401	K650TTA	Dennis DS155	John Dennis	WrL	1992	11 Lynton
5402	K651TTA	Dennis DS155	John Dennis	WrL	1992	21 Brixham
5403	K652TTA	Dennis DS155	John Dennis	WrL	1992	12 North Tawton
5404	K653TTA	Dennis DS155	John Dennis	WrL	1992	41 Ottery St Mary
5405	K654TTA	Dennis DS155	John Dennis	WrL	1992	35 Bampton
5439	YCV942T	Land Rover Series III 109	Carmichael Redwing	L4P	1979	54 Kingston Volunteers
5467	L311DOD	Dennis Rapier TF203	John Dennis	WrL	1994	50 Greenbank
5468	L312DOD	Dennis Rapier TF203	John Dennis	WrL	1994	48 Camels Head
5469	L313DOD	Dennis Rapier TF203	John Dennis	WrL	1994	32 Exeter
5470	M556HOD	Mercedes-Benz 1124F	Carmichael International	WrL	1994	49 Crownhill
5471	M557HOD	Mercedes-Benz 1124F	Carmichael International	WrL	1994	01 Barnstaple
5472	M558HOD	Mercedes-Benz 1124F	Carmichael International	WrL	1994	47 Plymouth
5473	M559HOD	Mercedes-Benz 1124F	Carmichael International	WrL	1994	01 Barnstaple
5475	L626DOD	Iveco-Magirus 140-25A	GB Fire/Magirus DL30	TL	1994	49 Crownhill

Six Mercedes-Benz appliances feature in the Devon Fire & Rescue Service fleet. Four are standard 1124Fs, one of which was illustrated in the last edition. The other two are compact Water Tenders by Carmichael International upon the all wheel drive 917AF chassis. Fleet number 5477 (M554HOD) is one of this pair, it was issued to the retained firefighters at Lynton. *Robert Smith*

5477	M554HOD	Mercedes-Benz 917F	Carmichael International	WrT	1995	11 Lynton
5478	M561HOD	Mercedes-Benz 917F	Carmichael International	WrT	1995	56 Princetown
5480	Boat	Dell Quay Rigid Raider	Boat & Trailer	IRBt	1989	51 Plymstock
5481	M53KFJ	MAN L2000 10.224F	Saxon Sanbec	C/WrL	1995	03 Appledore
5482	M59KFJ	MAN L2000 10.224F	Saxon Sanbec	C/WrL	1995	07 Combe Martin
5483	M52KFJ	MAN L2000 10.224F	Saxon Sanbec	C/WrL	1995	16 Woolacombe
5484	M54KFJ	MAN L2000 10.224F	Saxon Sanbec	C/WrL	1995	42 Seaton
5485	M58KFJ	MAN L2000 10.224F	Saxon Sanbec	C/WrL	1995	29 Salcombe
5486	M62KFJ	MAN L2000 10.224F	Saxon Sanbec	C/WrL	1995	52 Bere Alston
5487	M65KFJ	MAN L2000 10.224F	Saxon Sanbec	C/WrL	1995	37 Colyton
5488	M61KFJ	MAN L2000 10.224F	Saxon Sanbec	C/WrL	1995	06 Chumleigh
5489	M68KFJ	MAN L2000 10.224F	Saxon Sanbec	C/WrL	1995	08 Hartland
5490	M73KFJ	MAN L2000 10.224F	Saxon Sanbec	C/WrL	1995	23 Chagford
5491	M74KFJ	MAN L2000 10.224F	Saxon Sanbec	C/WrL	1995	36 Budleigh Salterton
5492	M67KFJ	MAN L2000 10.224F	Saxon Sanbec	C/WrL	1995	22 Buckfastleigh
5493	M69KFJ	MAN L2000 10.224F	Saxon Sanbec	C/WrL	1995	55 Modbury
5494	M51KFJ	MAN L2000 10.224F	Saxon Sanbec	C/WrL	1995	19 Ashburton
5495	M63KFJ	MAN L2000 10.224F	Saxon Sanbec	C/WrL	1995	14 South Molton
5496	M56KFJ	MAN L2000 10.224F	Saxon Sanbec	C/WrL	1995	53 Ivybridge
5497	M64KFJ	MAN L2000 10.224F	Saxon Sanbec	C/WrL	1995	26 Kingsbridge
5498	M71KFJ	MAN L2000 10.224F	Saxon Sanbec	C/WrL	1995	15 Torrington
5499	SRL773Y	Land Rover 109	Rover	L4T	1996	48 Camels Head
5500	N419RJF	MAN M90 17.232	Penman Multilift	PM	1996	Headquarters
5501	Pod	Penman Multilift	BA Training Unit	BAt	1995	Headquarters
5502	Pod	Penman Multilift	BA Training Unit	BAt	1989	Headquarters
5514	P907YTT	Dennis Sabre TSD233	John Dennis	WrT	1996	32 Exeter
5525	P919YTT	MAN M2000	Saxon Sanbec	WrL	1996	51 Plymstock
5526	P933YTT	Mercedes-Benz 814DA	Mercedes-Benz	LP	1996	13 Okehampton
5527	P934YTT	MAN L2000 10.224F	Saxon Sanbec	C/WrL	1996	05 Braunton
5528	P935YTT	MAN L2000 10.224F	Saxon Sanbec	C/WrL	1996	09 Hatherleigh
5529	P936YTT	MAN L2000 10.224F	Saxon Sanbec	C/WrL	1996	10 Holsworthy
5530	P937YTT	MAN L2000 10.224F	Saxon Sanbec	C/WrL	1996	25 Dawlish
5531	P938YTT	MAN L2000 10.224F	Saxon Sanbec	C/WrL	1996	34 Axminster
5532	P939YTT	MAN L2000 10.224F	Saxon Sanbec	C/WrL	1996	46 Witheridge
5535	P940YTT	MAN 18.224	Penman Multilift	PM	1996	Headquarters

Notes: 5168 Was converted from a water ladder (WrL) to a reserve ET, in 1993; 5439 Was acquired from Cornwall County Fire Brigade in 1993; 5499 Was acquired from Cornwall County Fire Brigade in 1996.

DORSET FIRE BRIGADE

Dorset Fire Brigade, County Hall, Dorchester DT1 1XJ

AFX592T	Bedford TKG	HCB-Angus	WrT	1978	Reserve
BPR761T	Land Rover series III 109	Dorset FB	L4T	1979	Training Centre
BPR763T	Land Rover series III 109	Dorset FB	L4T	1979	A04 Beaminster
CJT161T	Land Rover series III 109	Dorset FB	BTU	1979	B18 Poole Quay
KJT214W	Land Rover series III 109	Dorset FB	L4T	1980	A05 Maiden Newton
WJT240X	Bedford TKG	HCB-Angus	WrT	1982	Reserve
WJT241X	Bedford TKG	HCB-Angus	WrT	1982	Driving School
WJT243X	Bedford TKG	HCB-Angus	WrT	1982	Reserve
WJT244X	Bedford TKG	HCB-Angus	WrT	1982	Driving School
WJT651X	Dodge S56	Corvesgate	CU	1982	B17 Hamworthy
BPR181Y	Bedford TKG	HCB-Angus	WrT	1982	A09 Sherborne
BPR183Y	Bedford TK1260	HCB-Angus	WrL	1982	Workshops
A937KJT	Bedford TK1260	HCB-Angus	WrL	1984	B16 Swanage
A938KJT	Bedford TK1260	HCB-Angus	WrL	1984	Reserve
A939KJT	Bedford TK1260	HCB-Angus	WrL	1984	A13 Blandford
A940KJT	Bedford TK1260	HCB-Angus	WrL	1984	A06 Portland
A108NEL	Dodge 50	Locomotors	FoT	1984	B18 Poole
B467TEL	Bedford TK1260	HCB-Angus	WrL	1985	Reserve
B468TEL	Bedford TK1260	HCB-Angus	WrL	1985	A08 Dorchester
C83XLJ	Land Rover 110	Dorset FB	L4T	1985	A12 Shaftesbury
C84XLJ	Land Rover 110	Dorset FB	L4T	1985	B20 Cranborne
C625XPR	Bedford TL	HCB-Angus	WrT	1985	Training Centre
D240ELJ	Volvo FL6.16	Angloco / Metz DL30	TL	1986	Reserve
D566FJT	Land Rover 110 6x6	Reynolds Boughton/FSE	RT	1986	Reserve
D159HEL	Land Rover 110	Dorset FB	L4T	1987	A07 Weymouth
E50MEL	Volvo FL6.16	Dennis / Simon Snorkel SS263	HP	1987	B23 Westbourne, Bournemouth
E999MJT	Volvo FL6.14	HCB-Angus	WrL	1987	A11 Gillingham
E568NRU	Volvo FL6.14	HCB-Angus	WrT	1987	Training Centre
E569NRU	Volvo FL6.14	HCB-Angus	WrL	1987	A05 Maiden Newton
E336OJT	Volvo FL6.14	HCB-Angus	WrL	1988	A07 Weymouth
E367PFX	Land Rover 110 6x6	HCB-Angus	L6P	1988	A01 Lyme Regis
F388NLC	Mercedes-Benz 917AF	HCB-Angus	MWrT	1988	B19 Wimborne
F338VFX	Land Rover 110 6x6	HCB-Angus	L6P	1988	A10 Sturminster Newton
F339VFX	Land Rover 110 6x6	HCB-Angus	L6P	1988	A11 Gillingham
F153XFX	Volvo FL6.14	HCB-Angus	WrL	1989	B26 Christchurch
F154XFX	Volvo FL6.14	HCB-Angus	WrL	1989	B15 Wareham
F155XFX	Volvo FL6.14	HCB-Angus	WrL	1989	B29 Verwood
G782CLJ	Land Rover 110	Dorset FB	L4T	1989	B29 Verwood
G95EEL	Land Rover 110	Dorset FB	L4T	1990	A02 Charmouth
G749ELJ	Volvo FL6.14	HCB-Angus	WrL	1990	A10 Sturminster Newton
G750ELJ	Volvo FL6.14	HCB-Angus	WrL	1990	B14 Bere Regis
G220FFX	Volvo FL6.17	Saxon / Simon Snorkel STS240	HP	1990	A07 Weymouth
H380KRU	Land Rover Defender 110	Dorset FB	L4T	1990	A08 Dorchester
H381KRU	Land Rover 110	Dorset FB	L4T	1990	A02 Charmouth
H627KJT	Volvo FL6.14	HCB-Angus	WrL	1990	B16 Swanage
H628KJT	Volvo FL6.14	HCB-Angus	WrL	1990	A01 Lyme Regis
H629KJT	Volvo FL6.14	HCB-Angus	WrL	1990	A04 Beaminster
H676KLJ	Vauxhall Astravan	Vauxhall/Dorset FB	EST	1990	A09 Sherborne
H677KLJ	Vauxhall Astravan	Vauxhall/Dorset FB	EST	1990	A06 Portland
H703LLJ	Mercedes-Benz 917AF	HCB-Angus	MWrT	1991	B17 Hamworthy
H704LLJ	Mercedes-Benz 917AF	HCB-Angus	MWrT	1991	B21 Ferndown
H695LRU	Volvo FL6.14	HCB-Angus	WrL	1991	A12 Shaftesbury
H696LRU	Volvo FL6.14	HCB-Angus	WrL	1991	A03 Bridport
H697LRU	Volvo FL6.14	HCB-Angus	WrL	1991	B17 Hamworthy
H698LRU	Volvo FL6.14	HCB-Angus	WrL	1991	B22 Redhill Park, Bournemouth
H470NEL	Volvo FL6.14	HCB-Angus	WrL	1991	Reserve
H471NEL	Volvo FL6.14	HCB-Angus	WrL	1991	A06 Portland
J86ORU	Volvo FL6.14	HCB-Angus	WrL	1991	B23 Westbourne, Bournemouth
J87ORU	Volvo FL6.14	HCB-Angus	WrL	1991	B18 Poole
J89ORU	Volvo FL6.14	HCB-Angus	WrL	1991	B20 Cranborne
J491PPR	Land Rover Defender 110	Dorset FB	L4T	1991	B18 Poole
J492PPR	Land Rover Defender 110	Dorset FB	L4T	1991	B14 Bere Regis

The Fire Brigade Handbook

The majority of Dorset Fire Brigade's pumping appliances are Water Tender Ladders based on Volvo FL6 chassis. Two appliances, however, have the more powerful FS7 chassis and were finished as combined Water Tender Ladder/Emergency Tenders by HCB-Angus. L344CEL is the newer of the pair and is on the run at Poole. *Robert Smith*

K143VJT	Volvo FL6.14	HCB-Angus	WrL	1993	A07 Weymouth
K144VJT	Volvo FL6.14	HCB-Angus	WrL	1993	B21 Ferndown
K907VRU	Volvo FL6.18	Wincanton/Reynolds Boughton	WrC	1993	B15 Wareham
K473UJT	Land Rover Defender 110	Dorset FB	L4T	1993	A13 Blandford
K474UJT	Land Rover Defender 110	Dorset FB	L4T	1993	B17 Hamworthy
K475UJT	Land Rover Defender 110	Dorset FB	L4T	1993	B26 Christchurch
K476UJT	Land Rover Defender 110	Dorset FB	L4T	1993	B19 Wimborne
L697AFX	Volvo FS7.18	HCB-Angus	WrL/ET	1993	A07 Weymouth
L727BFX	Mercedes-Benz 917AF	HCB-Angus	MWrT	1993	B15 Wareham
L728BFX	Leyland-DAF Freighter 45-130	Locomotors	BAT	1993	B26 Christchurch
L729BFX	Leyland-DAF Freighter 45-130	Locomotors	BAT	1993	A07 Weymouth
L730BFX	Volvo FL6.14	HCB-Angus	WrL	1993	B24 Springbourne
L339CEL	Volvo FL6.14	HCB-Angus	WrL	1993	A08 Dorchester
L340CEL	Volvo FL6.14	HCB-Angus	WrL	1993	A09 Sherbourne
L343CEL	Mercedes-Benz 1124F	HCB-Angus	MWrT	1994	A03 Bridport
L344CEL	Volvo FS7.18	HCB-Angus	WrL/ET	1994	B18 Poole
L327CLJ	Land Rover Defender 110	Dorset FB	L4T	1994	B21 Ferndown
L328CLJ	Land Rover Defender 110	Dorset FB	L4T	1994	A12 Shaftesbury
L329CLJ	Land Rover Defender 110	Dorset FB	L4T	1994	B15 Wareham
M143LLJ	Volvo FL6.14	Saxon Sanbec	WrL	1995	B19 Wimbourne
M144LLJ	Volvo FL6.14	Saxon Sanbec	WrL	1995	B16 Swanage
M145LLJ	Volvo FL6.14	Saxon Sanbec	WrL	1995	A13 Blandford
N374NRU	Dennis Sabre TSD233	John Dennis	WrL	1995	B26 Christchurch
N375NRU	Dennis Sabre TSD233	John Dennis	WrL	1995	A03 Bridport
N455SPR	Dennis Sabre TSD233	John Dennis	WrL	1995	B18 Poole
N456SPR	Dennis Sabre TSD233	John Dennis	WrL	1996	
P	Dennis Sabre TSD233	John Dennis	WrL	1996	
P	Dennis Sabre TSD233	John Dennis	WrL	1996	
P	Dennis Sabre TSD233	John Dennis	WrL	1996	

Note:
F388NLC was formerly a demonstrator; H676/7KLJ were converted from ancilliary vans.

The brigade workshops of Dumfries & Galloway Fire Brigade converted two Renault Messenger panel vans for use as Incident Support Units in 1994. Upon completion they were placed on the run at Dumfries and Newton Stewart is L279VSW, the Dumfries example. *Robert Smith*

Scania appliances have gained popularity with four of the eight Scottish fire brigades, most of which had previously settled upon Dodge based machines. Two new P93M-220s with Emergency One bodywork and Angus-Sacol ladder gantries went on into service at Dumfries in 1995. The gantry can be seen in this view of M75BSM. *Alistair MacDonald*

DUMFRIES & GALLOWAY FIRE BRIGADE

Dumfries and Galloway Fire Brigade, Brooms Road, Dumfries DG1 2DZ

Reg	Chassis	Body	Type	Year	Station
YSM430W	Dodge G1613	Carmichael / Magirus DL30	TL	1981	D Reserve
EMS820Y	Dodge G12C	Mountain Range	WrL	1983	B Reserve
B721LSW	Dodge S56	Caledonian	GPV	1985	B6 Newton Stewart
B867LSW	Dodge G12C	Mountain Range	WrL	1984	A Reserve
B868LSW	Dodge G12C	Mountain Range	WrL	1984	A Reserve
C745OSW	Dodge S66	Angloco/Penman	BASU	1986	D1 Dumfries
D195SSW	Dodge G08	Excalibur CBK	C/FRT	1987	B5 New Galloway
D196SSW	Dodge G08	Excalibur CBK	C/FRT	1987	B4 Kirkcudbright
D197SSW	Dodge G13	Alexander	FRT	1987	B7 Stranraer
D198SSW	Dodge G13	Alexander	FRT	1986	A3 Langholm
D970SSW	Dodge G08	Excalibur CBK	FRT	1987	B Reserve
D971SSW	Dodge G08	Excalibur CBK	C/FRT	1987	A7 Thornhill
D972SSW	Dodge G08	Excalibur CBK	C/FRT	1987	B2 Dalbeattie
E277WSW	Renault-Dodge G08	Excalibur CBK	C/FRT	1987	B3 Gatehouse of Fleet
Trailer	Load Lugger/D&GFB	Equipment Trailer	EqT	1987	AV1 Eskdalemuir Volunteers
Trailer	Load Lugger/D&GFB	Equipment Trailer	EqT	1987	BV1 Drummore Volunteers
E682WSW	Renault-Dodge G08	Excalibur CBK	C/FRT	1988	B6 Newton Stewart
E683WSW	Renault-Dodge G08	Excalibur CBK	C/FRT	1988	B8 Whithorn
E684WSW	Renault-Dodge G08	Excalibur CBK	C/FRT	1988	A6 Sanquhar
E685WSW	Renault-Dodge G08	Excalibur CBK	C/FRT	1988	A1 Annan
E690WSW	Renault-Dodge G08	Scott	ICU	1988	D1 Dumfries
F894ASW	Renault-Dodge G13	Penman Multilift	PM	1989	D1 Dumfries
Pod	Penman	BA Taining Unit	BAT	1989	D1 Dumfries
F897ASW	Renault-Dodge G13	Excalibur CBK	FRT	1989	A2 Gretna
F548CSM	Ford Escort	Ford/D&GFB	FIU	1989	D1 Dumfries
G629ESM	Ford Transit	Bradleigh	PCV	1990	B6 Newton Stewart
G630ESM	Ford Transit	Bradleigh	PCV	1990	D1 Dumfries
G632ESM	Renault-Dodge G13	Excalibur CBK	FRT	1990	D1 Dumfries
H299HSW	Scania P93M-280	Angloco / Bronto Skylift 28-2Ti	ALP	1990	A5 Moffat
H466JSW	Scania G93M-120	Emergency One	WrL	1991	A5 Moffat
H467JSW	Scania G93M-210	Emergency One	FRT	1991	A4 Lockerbie
H246KSM	Scania G93M-210	Emergency One	WrT	1991	B1 Castle Douglas
K53TSM	Scania G93M-210	Emergency One	FRT	1993	B6 Newton Stewart
K54TSM	Scania G93M-210	Emergency One	FRT	1993	B7 Stranraer
L279VSW	Renault Messenger B110	D&GFB	ISU	1994	D1 Dumfries
L281VSW	Renault Messenger B110	D&GFB	ISU	1994	B6 Newton Stewart
L612VSW	Scania P93M-220	Emergency One	FRT	1994	A1 Annan
M75BSM	Scania P93M-220	Emergency One	FRT	1995	D1 Dumfries
M76BSM	Scania P93M-220	Emergency One	FRT	1995	D1 Dumfries

Note: C745PSW was previously a CIU

DURHAM COUNTY FIRE & RESCUE BRIGADE

Durham County Fire & Rescue Brigade, Framwellgate Moor, Durham DH1 5JR

004	N430RCN	Dennis Rapier TF203	Emergency One	RWL	1995	G Durham
005	N431RCN	Dennis Rapier TF203	Emergency One	RWL	1995	Q Darlington
006	M837KBB	Dennis Rapier TF203	Carmichael International	RWL	1995	M Newton Aycliffe
007	M838KBB	Dennis Rapier TF203	Carmichael International	RWL	1995	N Bishop Auckland
008	M839KBB	Dennis Rapier TF203	Carmichael International	RWL	1995	B Stanley
012	K492PCN	Dennis Rapier TF202	Carmichael International	RWL	1992	A Consett
013	K493PCN	Dennis Rapier TF202	Carmichael International	RWL	1992	P Barnard Castle
014	K494PCN	Dennis Rapier TF202	Carmichael International	RWL	1992	D Seaham
015	K495PCN	Dennis Rapier TF202	Carmichael International	RWL	1992	I Stanhope
016	J763KNL	Dennis Rapier TF202	Carmichael International	RWL	1992	G Durham
017	J764KNL	Dennis Rapier TF202	Carmichael International	RWL	1992	C Fencehouses
018	J765KNL	Dennis Rapier TF202	Carmichael International	RWL	1992	Q Darlington
019	J766KNL	Dennis Rapier TF202	Carmichael International	RWL	1992	E Peterlee
020	H684ATN	Dennis DS153	Carmichael	WrL	1990	E Peterlee
021	H685ATN	Dennis DS153	Carmichael	WrL	1991	Q Darlington
022	H686ATN	Dennis DS153	Carmichael	WrL	1990	D Seaham
023	H687ATN	Dennis DS153	Carmichael	WrL	1991	G Durham
024	G633TJR	Dennis DS153	Carmichael	WrL	1990	M Newton Aycliffe
025	G634TJR	Dennis DS153	Carmichael	WrL	1990	C Fencehouses
026	G635TJR	Dennis DS153	Carmichael	WrL	1989	N Bishop Auckland
027	F377KCN	Dennis DS153	Fulton & Wylie	WrL	1989	K Spennymoor
028	F378KCN	Dennis DS153	Fulton & Wylie	WrL	1989	G Durham
029	E247DFT	Dennis DS153	Carmichael	WrL	1987	J Crook
030	E248DFT	Dennis DS153	Carmichael	WrL	1987	B Stanley
031	D177KFT	Dennis DS151	Carmichael	WrL	1987	I Stanhope
032	D178KFT	Dennis DS151	Carmichael	WrL	1987	L Sedgefield

Delivered during 1995, Durham County's 084, M415FRG is the brigades new heavy rescue vehicle witha boom by Boniface mounted upon a Volvo FL10 chassis. Allocated to Bishop Auckland, it replaced a Dennis Maxim/Holmes machine some twenty five years its senior. *Keith Grimes*

Durham County Fire And Rescue Brigade have, in recent years, taken Dennis Rapier chassis for pumping appliances and Volvo products for its specialist heavy duty roles. Representing the Volvo range is 085, N432RCN, the brigades first aerial ladder-platform and seen here on a FL10 chassis. It was photographed at Darlington where training on the appliance was being undertaken. *Keith Grimes*

033	C833LTN	Dennis DS151	Dennis	WrL	1985	F Wheatley Hill
034	C834LTN	Dennis DS151	Dennis	WrL	1985	I Stanhope
035	C835LTN	Dennis DS151	Dennis	WrL	1985	O Middleton-in-Teesdale
036	C836LTN	Dennis DS151	Dennis	WrL	1985	Reserve
052	TPT17V	Dennis D/Rolls Royce	Dennis	WrL	1979	Reserve
065	B210GNL	Dennis DS151	Dennis	WrL	1985	J Crook
066	FBR222Y	Dennis SS133	Dennis	WrL	1983	A Consett
067	FBR223Y	Dennis SS133	Dennis	WrL	1983	P Barnard Castle
068	FBR224Y	Dennis SS133	Dennis	WrL	1983	P Barnard Castle
069	KCU866X	Dennis D/Perkins	Dennis	WrL	1981	Training Centre
070	KCU867X	Dennis D/Perkins	Dennis	WrL	1981	Reserve
071	APT967W	Dennis D/Perkins	Dennis	WrL	1980	Training Centre
072	APT968W	Dennis D/Perkins	Dennis	WrL	1980	Reserve
081	L515XBB	Volvo FL6.17	Multilift	PM	1994	Training Centre
082	L516XBB	Dennis SS235	Devcoplan/DF&RB	SIU	1994	G Durham
083	L517XBB	Dennis SS235	John Dennis	FoT	1994	N Bishop Auckland
084	M415FRG	Volvo FL10	Boniface	HRV	1995	Q Darlington
085	N432RCN	Volvo FL10	Angloco/Bronto Skylift F32HDT	ALP	1996	Bishop Auckland
096	Pod	Penman	BA Training Unit	BAT	1994	Training Centre
097	Trailer	Tow-a-van/Newcastle Trailers	Control Unit	CU	1993	G Durham
098	Trailer	Tow-a-van/Newcastle Trailers	Lighting Unit	LU	1992	Q Darlington
099	Trailer	Margrove Caravans	Fire Prevention Unit	ExU	1987	G Durham
191	B648GTN	Dennis DFD133	Carmichael/Magirus DL30	TL	1985	G Durham
200	FBR221Y	Dennis DF133	Carmichael/Magirus DL30	TL	1983	Q Darlington
211	F376KCN	Renault-Dodge S46	Rootes	PCV	1988	Training Centre
213	H753ABB	Renault-Dodge S56	Rootes	GPV	1990	G Durham
219	F468OTN	Renault-Dodge S56	Devcoplan	WRu	1989	G Durham
224	K462TFT	Land Rover Defender 110 Tdi	Land Rover	L4V	1993	A Consett
230	J782KNL	Land Rover Defender 110 Tdi	Land Rover	L4V	1992	E Peterlee
232	J781KNL	Land Rover Defender 110 Tdi	Land Rover	L4V	1992	Q Darlington
234	H275CVK	Land Rover Defender 110 Tdi	Land Rover	L4V	1991	G Durham
235	N119OTN	Land Rover Defender 110 Tdi	Land Rover	L4V	1996	C Fencehouses
236	N120OTN	Land Rover Defender 110 Tdi	Land Rover	L4V	1996	M Newton Aycliffe

EAST SUSSEX COUNTY FIRE BRIGADE

East Sussex County Fire Brigade, 24 King Henry's Road, Lewes, East Sussex BN7 1BZ

SCD885R	ERF 84CS	ERF /Simon Snorkel SS263	HP	1977	02 Hove
LVH196T	Bedford YMT	Anglo/Yeates Catering	CU	1978	01 Lewes
YAP559T	Bedford TK860	Gardiner	GPV	1978	08 Hastings
PWV74X	Bedford KG	Pilcher Greene	ExU	1981	11 Eastbourne
RAP992X	Bedford KG	HCB-Angus CSV	WrL	1981	Training Centre
RCD659X	Bedford KG	HCB-Angus CSV	WrL	1981	Training Centre
RUF584X	Bedford KG	HCB-Angus CSV	WrL	1981	Reserve
A168CUF	Bedford KG	HCB-Angus HSC	WrL	1984	Reserve
A169CUF	Bedford KG	HCB-Angus HSC	WrL	1984	Reserve
A170CUF	Bedford KG	HCB-Angus HSC	WrL	1984	Reserve

Now the oldest operational appliance with East Sussex County Fire Brigade is SCD885R. This appliance is on the run at Hove and dates from 1977. It features a Cummins-engined ERF unit onto which a Simon Snorkel SS263 boom is mounted.
Gavin Stewart

A heavy rescue unit was allocated to Lewes in 1995. Volvo FL6, M128BPN, carries Bence bodywork for its specialist role, its equipment including a HIAB crane which is visible in this picture taken at Lewes fire station.
Gavin Stewart

A210CYJ	MAN 16.240FR	Angloco / Metz DLK23-12	TL	1984	22 Bexhill
A779EPN	Jeep J20 Ijene/Scothorn 6x4	Angloco	RT	1984	01 Lewes
B597GAP	Bedford M1120 4x4	Angloco/TICO T350WT	H4P	1984	13 Crowborough
C443OAP	Bedford KG	HCB-Angus HSC	WrL	1985	06 Newhaven
C444OAP	Bedford KG	HCB-Angus HSC	WrL	1985	07 Seaford
C445OAP	Bedford KG	HCB-Angus HSC	WrL	1985	17 Burwash
C446OAP	Bedford KG	HCB-Angus HSC	WrL	1985	18 Battle
C820PPN	Bedford TL	Gardiner/TICO 235	GPV	1986	09 The Ridge, Hastings
C821PPN	Bedford TL	Gardiner/TICO 235	GPV	1986	02 Hove
C945SAP	Bedford KG	HCB-Angus HSC	WrL	1986	21 Heathfield
C946SAP	Bedford KG	HCB-Angus HSC	WrL	1986	15 Wadhurst
C947SAP	Bedford KG	HCB-Angus HSC	WrL	1986	20 Rye
D455XAP	Bedford TL	HCB-Angus HSC	WrL	1987	01 Lewes
D456XAP	Bedford TL	HCB-Angus HSC	WrL	1987	Reserve
D457XAP	Bedford TL	HCB-Angus HSC	WrL	1987	Reserve
D458XAP	Bedford TL	HCB-Angus HSC	WrL	1987	22 Bexhill
E328BAP	Volvo FL6.14	HCB-Angus	WrL	1987	21 Heathfield
E329BAP	Volvo FL6.14	HCB-Angus	WrL	1987	23 Pevensey
E184BPN	MAN 16.240FR	Angloco / Metz DLK30	TL	1987	11 Eastbourne
E741CYJ	Volvo FL6.14	HCB-Angus	WrL	1988	Training Centre
E742CYJ	Volvo FL6.14	HCB-Angus	WrL	1988	03 Barcombe
G408RAP	MAN 16.240FDS	Angloco / Metz DLK30	TL	1989	08 Hastings
G117TAP	GMC 3500 TV31403 4x4	Macclesfield Motor Bodies	OSU	1990	22 Bexhill
G118TAP	GMC 3500 TV31403 4x4	Macclesfield Motor Bodies	OSU	1990	12 Uckfield
G119TAP	GMC 3500 TV31403 4x4	Macclesfield Motor Bodies	OSU	1990	18 Battle
G121TAP	Volvo FL6.14	HCB-Angus	WrL	1990	14 Mayfield
G122TAP	Volvo FL6.14	HCB-Angus	WrL	1990	12 Uckfield
G123TAP	Volvo FL6.14	HCB-Angus	WrL	1990	13 Crowborough
G124TAP	Volvo FL6.14	HCB-Angus	WrL	1990	19 Broad Oak
G125TAP	Volvo FL6.14	HCB-Angus	WrL	1990	25 Hailsham
G126TAP	Volvo FL6.14	HCB-Angus	WrL	1990	07 Seaford
G127TAP	Volvo FL6.14	HCB-Angus	WrL	1990	20 Rye
G128TAP	Volvo FL6.14	HCB-Angus	WrL	1990	09 The Ridge
G129TAP	Volvo FL6.14	HCB-Angus	WrL	1990	15 Wadhurst
H521YAP	Mercedes-Benz 917AF	Reynolds Boughton/HIAB 050	FoT	1990	06 Newhaven
J974FWV	Volvo FL6.14	Excalibur CBK	WrL	1992	06 Newhaven
J975FWV	Volvo FL6.14	Excalibur CBK	WrL	1992	10 Forest Row
J976FWV	Volvo FL6.14	Excalibur CBK	WrL	1992	18 Battle
K706LYJ	Volvo FS7.18	Maidment (1993)	WrC	1993	13 Crowborough
K707LYJ	Mercedes-Benz 1726	Angloco/Metz DLK30PLC	TL	1993	04 Preston Circus, Brighton
K708LYJ	Volvo FL6.14	Excalibur CBK	WrL	1993	22 Bexhill
K709LYJ	Volvo FL6.14	Excalibur CBK	WrL	1993	01 Lewes
K710LYJ	Volvo FL6.14	Excalibur CBK	WrL	1993	24 Herstmonceux
L249SPN	Volvo FL6.14	Excalibur CBK	WrL	1994	12 Uckfield
L250SPN	Volvo FL6.14	Excalibur CBK	WrL	1994	13 Crowborough
L251SPN	Volvo FL6.14	Excalibur CBK	WrL	1994	11 Eastbourne
L252SPN	Volvo FL6.14	Excalibur CBK	WrL	1994	02 Hove
M126BPN	Volvo FL6.14	Excalibur CBK	WrL	1994	04 Preston Circus, Brighton
M127BPN	Volvo FL6.14	Excalibur CBK	WrL	1995	09 The Ridge, Hastings
M128BPN	Volvo FL7.18	W H Bence/HIAB	HRV	1995	01 Lewes
M732CBX	Renault Trafic T35D	Bott	BASU	1995	20 Rye
M733CBX	Renault Trafic T35D	Bott	BASU	1995	Roedean, Brighton
M849FNJ	Ford Transit VE6	Ford	PCV	1995	02 Hove
M253FAP	Ford Transit VE6	Ford	PCV	1995	08 Hastings
N43PPN	Ford Transit VE6	Ford	PCV	1995	50 Maresfield
N951NUF	Ford Transit VE6	Ford	CSU	1996	11 Eastbourne
N748LUF	Volvo FL6.14	Excalibur CBK	WrL	1996	08 Hastings
N749LUF	Volvo FL6.14	Excalibur CBK	WrL	1996	05 Roedean, Brighton
N750LUF	Volvo FL6.14	Excalibur CBK	WrL	1996	04 Preston Circus, Brighton
N751LUF	Volvo FL6.14	Excalibur CBK	WrL	1996	02 Hove
N803BJK	Land Rover Defender	Rover	RRU	1996	22 Bexhill

Note:
PWV74X was originally a CIU; H521YAP was originally an Animal Rescue Unit.

ESSEX COUNTY FIRE & RESCUE SERVICE

Essex County Fire & Rescue Service, Rayleigh Close, Rayleigh Road, Hutton, Brentwood CM13 1AL

HNO247B	AEC Regent V	Merryweather	TL	1963	Miscellaneous
RVW836W	Bedford TKG	Benson	CU	1980	Headquarters
A346MVX	Dennis RS133	Dennis	RP	1984	Community Education Unit
A347MVX	Dennis RS133	Dennis	RP	1984	Training Centre
A348MVX	Dennis RS133	Dennis	RP	1984	Reserve
A350MVX	Dennis RS133	Dennis	WrL	1984	Reserve
A352MVX	Bedford M1160 4x4	Essex FB(1962)	HL	1984	E35 Hadleigh
B89BJN	Dennis RS135	Dennis	RP	1985	E22 West Mersea
B91BJN	Dennis RS135	Dennis	RP	1985	Reserve
B92BJN	Dennis RS135	Dennis	WrT	1985	W81 Halstead
B93BJN	Dennis RS135	Dennis	RP	1985	Reserve
B94BJN	Dennis RS135	Dennis	RP	1985	Reserve
B95BJN	Dennis RS135	Dennis	RP	1985	Reserve
B96BJN	Dennis RS135	Dennis	RP	1985	Reserve
B97BJN	Dennis RS135	Dennis	RP	1985	Reserve
B98BJN	Dennis RS135	Dennis	RP	1985	Training Centre
B99BJN	Dennis SS135	Saxon Sanbec	RT	1985	Reserve
C832GNO	Dennis RS135	Carmichael	RP	1986	Reserve
C833GNO	Dennis RS135	Carmichael	RP	1986	Reserve
C834GNO	Dennis RS135	Carmichael	RP	1986	E25 Witham
C835GNO	Dennis RS135	Carmichael	RP	1986	Reserve
C836GNO	Dennis RS135	Carmichael	WrT	1986	E18 Frinton
C837GNO	Dennis RS135	Carmichael	RP	1986	W86 Thaxted
C838GNO	Dennis RS135	Carmichael	RP	1986	E24 Coggeshall
C839GNO	Dennis RS135	Carmichael	RP	1986	Reserve
D886POO	Dennis DFS135	G & T Fire Control	DTV	1987	Driving School
D887POO	Dennis RS135	John Dennis	RP	1987	E21 Wivenhoe
D888POO	Dennis RS135	John Dennis	RP	1987	Reserve
D889POO	Dennis RS135	John Dennis	RP	1987	Training Centre
D890POO	Dennis RS135	John Dennis	RP	1987	Reserve
D891POO	Dennis RS135	Carmichael	RP	1987	E47 Hawkwell
D892POO	Dennis DF135	Carmichael	FoT	1987	W70 Harlow Central
D893POO	Dennis RS135	John Dennis	WrT	1987	E25 Witham
D894POO	Dennis RS135	John Dennis	RP	1987	W89 Epping
D895POO	Dennis RS135	John Dennis	RP	1987	W78 Braintree

The new Foam Carrier appliance in Essex is based at Grays fire station. This Scania G93H-280 carries paletised foam compound and associated equipment. At incidents this is usually unloaded by a Moffett Mounty all terrain fork lift truck suspended at the rear of the truck.
Edmund Gray

For heavy duty roles such as aerial appliances, Essex have procured Scania chassis. Four aerial platform appliances with Simon Snorkel ST300-S 30-metre booms and Saxon bodywork are mounted onto the powerful P113H-310 chassis. Based at Colchester, J114UPU illustrates the type. A new aerial appliance is due shortly and this machine will feature the new ST340-S booms and will be among the tallest aerial appliances in the UK at around 34 metres. *Edmund Gray*

D896POO	Dennis RS135	John Dennis	RP	1987	Reserve
D897POO	Dennis RS135	John Dennis	RP	1987	E48 Rayleigh
E875ANO	Iveco 60.10	Iveco/Essex CF&RS	BAT	1988	Headquarters (Hutton)
E876ANO	Dennis RS135	John Dennis	WrT	1988	E11 Dovercourt
E877ANO	Dennis RS135	John Dennis	RP	1988	Training Centre
E878ANO	Dennis RS135	John Dennis	RP	1988	E11 Dovercourt
E879ANO	Dennis RS135	John Dennis	RP	1988	E46 Maldon
E880ANO	Dennis RS135	John Dennis	RP	1988	Reserve
E881ANO	Dennis RS135	John Dennis	RP	1988	Reserve
E882ANO	Dennis RS135	John Dennis	RP	1988	E23 Tiptree
E883ANO	Dennis RS135	John Dennis	RP	1988	Reserve
E884ANO	Dennis RS135	John Dennis	RP	1988	W83 Stanstead
E885ANO	Dennis DF135	Carmichael	FoT	1987	W52 Basildon
F921NHK	Dennis RS135	John Dennis	RP	1989	E43 Burnham
F922NHK	Dennis RS135	John Dennis	WrT	1989	Driving School
F923NHK	Dennis RS135	John Dennis	RP	1989	E20 Brightlingsea
F924NHK	Dennis RS135	John Dennis	RP	1989	E11 Dovercourt
F925NHK	Dennis RS135	John Dennis	RP	1989	Reserve
F926NHK	Dennis RS135	John Dennis	RP	1989	Driving School
F927NHK	Dennis RS135	John Dennis	RP	1989	W80 Sible Hedingham
F928NHK	Mercedes-Benz Unimog 1300L	John Dennis	L/WrT	1989	W83 Stanstead
G73WWC	Dennis RS135	John Dennis	RP	1990	E18 Frinton
G74WWC	Dennis RS135	John Dennis	RP	1990	W68 Billericay
G75WWC	Dennis RS135	John Dennis	WrT	1990	E35 Hadleigh
G76WWC	Dennis RS135	John Dennis	RP	1990	W79 Wethersfield
G77WWC	Dennis RS135	John Dennis	WrT	1990	E46 Maldon
G78WWC	Dennis RS135	John Dennis	RP	1990	E45 Tollesbury
G79WWC	Dennis RS135	John Dennis	RP	1990	W85 Saffron Walden
G80WWC	Dennis RS135	John Dennis	RP	1990	E49 Rochford
G81WWC	Dennis RS135	John Dennis	RP	1990	E17 Manningtree
H788GAR	Iveco-Ford Cargo 1215	John Dennis (1995)	HL	1991	W68 Billericay
H276HVX	Dennis SS235	John Dennis	RT	1991	W70 Harlow Central

Three Scania G93M-25-based Rescue Tenders were delivered to Essex County Fire And Rescue service in 1994, the rear end of the appliances were designed to enable Moffett Mounty all terrain fork lift trucks to be transported. These views show L675LOO at Grays fire station. The large sliding shutter at the rear of the Leicester Carriage built body houses a pallet containing Hydraulic Cutting and Spreading gear and this is transported by the Fork Lift to the closest point at incidents. *Keith Grimes*

Reg	Chassis	Body	Type	Year	Station
H277HVX	Scania P113H-310	Saxon Sanbec/Simon ST300-S	ALP	1991	E34 Chelmsford
H278HVX	Dennis RS235	John Dennis	WrT	1991	W78 Braintree
H279HVX	Dennis RS235	John Dennis	RP	1991	W82 Old Harlow
H281HVX	Dennis RS235	John Dennis	WrT	1991	W66 Basildon
H282HVX	Dennis RS235	John Dennis	RP	1991	W67 Ingatestone
H283HVX	Dennis RS235	John Dennis	WrT	1991	W51 Brentwood
H284HVX	Dennis RS235	John Dennis	WrT	1991	E30 Southend
H285HVX	Dennis RS235	John Dennis	RP	1991	W84 Newport
H576MEV	Steyr Pinzgauer D718K	Carmichael	L/WrT	1991	W87 Dunmow
J114UPU	Scania P113H-310	Saxon Sanbec/Simon ST300-S	ALP	1991	E10 Colchester
J115UPU	Steyr Pinzgauer D718K	Saxon Sanbec	L/WrT	1992	E17 Manningtree
J116UPU	Dennis RS237	John Dennis	RP	1992	W66 Corringham
J117UPU	Dennis RS237	John Dennis	RP	1992	Driving School
J118UPU	Dennis RS237	John Dennis	WrT	1992	Driving School
J119UPU	Dennis RS237	John Dennis	RP	1992	W70 Harlow Central
J120UPU	Dennis RS237	John Dennis	RP	1992	W81 Halstead
J121UPU	Dennis RS237	John Dennis	RP	1992	W69 Wickford
J122UPU	Dennis RS237	John Dennis	RP	1992	W88 Leader Roding
J754AAR	Scania P113H-310	Saxon Sanbec/Simon ST300-S	ALP	1992	E30 Southend
K385DOO	Scania P92ML-230	Leicester Carriage	CU	1993	Headquarters (Hutton)
K386DOO	Dennis RS237	John Dennis	WrT	1993	W54 Canvey Island
K387DOO	Dennis RS237	John Dennis	WrT	1993	W55 Tilbury
K388DOO	Dennis RS237	John Dennis	WrT	1993	W70 Harlow Central
K389DOO	Dennis Rapier TF202	John Dennis	WrT	1993	W50 Grays
K390DOO	Dennis Rapier TF202	John Dennis	RP	1993	E10 Colchester
K391DOO	Dennis Rapier TF202	John Dennis	RP	1993	E42 Shoeburyness
L663LOO	Scania P113H-310	Saxon Sanbec/Simon ST300-S	ALP	1991	W52 Basildon
L664LOO	Steyr-Daimler-Puch Pinzgauer	John Dennis	L/WrT	1994	W85 Saffron Walden
L665LOO	Steyr-Daimler-Puch Pinzgauer	John Dennis	L/WrT	1994	E43 Burnham
L667LOO	Dennis RS241	John Dennis	WrT	1993	E34 Chelmsford
L668LOO	Dennis RS241	John Dennis	WrT	1993	W52 Basildon
L669LOO	Dennis RS241	John Dennis	RP	1993	E12 Clacton
L671LOO	Dennis Rapier TF203	John Dennis	RP	1994	E44 Tillingham
L672LOO	Dennis Rapier TF203	John Dennis	RP	1994	W87 Dunmow
L673LOO	Dennis Rapier TF203	John Dennis	RP	1994	W50 Grays
L674LOO	Dennis Rapier TF203	John Dennis	WrT	1994	E10 Colchester
L675LOO	Scania G93M-250	Leicester Carriage	RT	1994	W50 Grays
L676LOO	Scania G93M-250	Leicester Carriage	RT	1994	E10 Colchester
L677LOO	Scania G93M-250	Leicester Carriage	RT	1994	E31 Leigh-on-Sea
M539UNO	Scania G93M-280	Leicester Carriage	FoT	1995	W50 Grays
M540UNO	Dennis RS237	John Dennis	RP	1994	W72 Loughton
N35COO	Moffett Mounty M2403N	Normally carried by L675LOO	FLT	1995	W50 Grays
N36COO	Moffett Mounty M2403N	Normally carried by L676LOO	FLT	1995	E10 Colchester
N37COO	Moffett Mounty M2403N	Normally carried by L677LOO	FLT	1995	E31 Leigh on Sea
N402CPU	Moffett Mounty M2403N	Normally carried by M539UNO	FLT	1995	W50 Grays
N461BVW	Dennis Rapier TF203	John Dennis	RP	1996	W51 Brentwood
N462BVW	Dennis Rapier TF203	John Dennis	RP	1996	E34 Chelmsford
N463BVW	Dennis Rapier TF203	John Dennis	WrT	1996	E12 Clacton
N464BVW	Dennis Rapier TF203	John Dennis	RP	1996	E31 Leigh on Sea
N465BVW	Dennis Sabre TSD233	John Dennis	RP	1996	W52 Basildon
N466BVW	Dennis Sabre TSD233	John Dennis	RP	1996	W73 Waltham Abbey
N467BVW	Dennis Sabre TSD233	John Dennis	RP	1996	E30 Southend
N468BVW	Dennis Sabre TSD233	John Dennis	RP	1996	W54 Canvey Island
N469BVW	Dennis Sabre TSD233	John Dennis	RP	1996	W71 Ongar
N470BVW	Dennis Sabre TSD233	John Dennis	RP	1996	E35 Hadleigh
N471BVW	Dennis Sabre TSD233	John Dennis	RP	1996	E32 South Woodham Ferrers
N472BVW	Dennis Sabre TSD233	John Dennis	RP	1996	E33 Great Baddow
N473BVW	Dennis Sabre TSD233	John Dennis	RP	1996	W55 Tilbury
N474BVW	Dennis Sabre TSD233	John Dennis	WrT	1996	W72 Loughton
N503GJN	Scania P113HL-320	Saxon/Simon Snorkel ALP340	ALP	1996	W70 Harlow Central
P	Dennis Sabre TSD233	John Dennis	WrT	1997	
P	Dennis Sabre TSD233	John Dennis	WrT	1997	
P	Dennis Sabre TSD233	John Dennis	WrT	1997	
P	Dennis Sabre TSD233	John Dennis	WrT	1997	
P	Dennis Sabre TSD233	John Dennis	WrT	1997	
P	Dennis Sabre TSD233	John Dennis	WrT	1997	
P	Dennis Sabre TSD233	John Dennis	WrT	1997	
P	Dennis Sabre TSD233	John Dennis	WrT	1997	

Note: D886POO was formerly a TL

The Fife Fire and Rescue service continue to take delivery of popular Volvo chassied appliances. M630RFS, from the 1994 delivery of Water Tenders, is based on the usual FL6 chassis. The substantial cab roof dome and tall sided bodywork are typical of fire appliances built by the Alexander factory in Belfast, part of the Mayflower group. This appliance is based at Rosyth. *Andrew Fenton*

Allocated to Rosythe is Fife's Meregency Support Unit, N576ASX. This vehicle is seen shortly after delivery and shows the Penman bodywork module mounted on a Volvo FL6 low profile chassis. *Andrew Fenton*

FIFE FIRE & RESCUE SERVICE

Fife Fire and Rescue Service, Strathore Road, Thornton, Kirkcaldy, Fife KY1 4DF

Reg	Chassis	Body	Type	Year	Location
BLS743V	Dodge G13	Fulton & Wylie	BAV	1980	Headquarters
EMS262V	Dodge G13	Methven	ET/ESU	1980	Reserve
ASG998W	Dodge G13	HCB-Angus	WrL	1981	Reserve
GFS75X	Leyland Terrier	Leyland	HL	1981	Headquarters
JSC473X	Dodge G13	Fulton & Wylie	WrL	1981	Reserve
JSC474X	Dodge G13	Fulton & Wylie	WrL	1981	Reserve
VFS111Y	Dodge G13	HCB-Angus	WrL	1983	Training Centre
VFS112Y	Dodge G13	HCB-Angus	WrL	1983	LGV Training Centre
B999MSF	Dodge G16	Fulton & Wylie/Simon SS263	HP	1985	A3 Kirkcaldy
C459RSF	Dodge G16	Fulton & Wylie/Simon SS263	HP	1985	A1 Dunfermline
C104VSC	Dodge G13	Fulton & Wylie	WrL	1986	Reserve
D415FSX	Dodge G10	Methven	CU	1986	Headquarters
D904ESG	Dodge G13	Alexander	WrL	1987	B7 St.Monans
D905ESG	Dodge G13	Alexander	WrL	1987	B5 Newburgh
E406MSX	Renault-Dodge G13	Alexander	WrL	1988	B3 Burnisland
E407MSX	Renault-Dodge G13	Alexander	WrL	1988	B8 Tayport
F538XSC	Volvo FL6.14	Alexander	WrL	1989	B6 St Andrews
F539XSC	Volvo FL6.14	Alexander	WrL	1989	B4 Cupar
G33HSC	Volvo FL6.14	Alexander	WrL	1990	B2 Auchtermuchty
G44HSC	Volvo FL6.14	Alexander	WrL	1990	B1 Anstruther
G583VNA	Renault-Dodge S56	Lynton	TU	1990	Community Edulation
Trailer	Lynton	Lynton	CEU	1990	Community Education
H81NSX	Volvo FL6.14	Alexander	WrL	1991	B6 St Andrews
H82NSX	Volvo FL6.14	Alexander	WrL	1991	B4 Cupar
J83CTS	Volvo FL6.14	Mountain Range	ET	1991	A4 Lochgelly
J649USF	Volvo FL6.14	Emergency One	WrL	1992	LGV Training Centre
J650USF	Volvo FL6.14	Emergency One	WrL	1992	A5 Methil
K961BSF	Volvo FL6.14	Alexander	WrL	1992	A3 Kirkcaldy
K962BSF	Volvo FL6.14	Alexander	WrL	1992	A3 Kirkcaldy
K479CSG	Renault-Dodge S56	Lynton/Outreach/Palfinger	TU	1993	A2 Kirkaldy
Trailer	Lynton	Lynton	CEU	1993	Community Education
L441HSC	Volvo FL6.14	Alexander	WrL	1993	A4 Lochgelly
L442HSC	Volvo FL6.14	Alexander	WrL	1993	A4 Lochgelly
L442HSC	Volvo FL6.14	Alexander	WrL	1993	A4 Lochgelly
M628RFS	Volvo FL6.14	Alexander	WrL	1994	A1 Dunfermline
M629RFS	Volvo FL6.14	Alexander	WrL	1994	A1 Dunfermline
M630RFS	Volvo FL6.14	Alexander	WrL	1994	A6 Rosyth
N573ASX	Volvo FL6.14	Alexander	WrL	1996	A2 Glenrothes
N574ASX	Volvo FL6.14	Alexander	WrL	1996	A2 Glenrothes
N575ASX	Volvo FL6.14	Alexander	WrL	1996	A5 Methil
N576ASX	Volvo FL6.14	Penman	ESU	1996	A6 Rosyth

Fife Fire & Rescue Service operate two similar Hydraulic Platform appliances based at Kirkaldy and Dunfermline. Shown during a training exercise is the Kirkaldy machine.
Fife Fire & Rescue Service

GLOUCESTERSHIRE FIRE & RESCUE SERVICE

Gloucestershire Fire & Rescue Service, Keynsham Road, Cheltenham,
Gloucestershire GL53 7PY

608	A210HFH	Dodge RG13C	Carmichael/Watts	PM	1984	05 Gloucester
610	A194LDG	Dodge RG13C	Carmichael/Watts	PM	1984	12 Cheltenham
611	TDA441Y	Dodge RG13C	Carmichael	GPV	1983	Withdrawn
623	B536MDF	Dodge RG13C	Carmichael	WrL	1984	Fire Safety
624	B537MDF	Dodge RG13C	Carmichael	WrL	1984	Withdrawn
625	B538MDF	Dodge RG13C	Carmichael	WrL	1984	Withdrawn
626	B633NDF	Dodge RG13C	Carmichael	WrL	1984	Training
630	B183SDF	Dodge RG13C	Carmichael/Watts	PM	1985	21 Cirencester
631	B184SDF	Dodge RG13C	Carmichael/Watts	PM	1985	Fire Safety
632	B185SDF	Dodge RG13C	Carmichael	WrL	1985	Training Centre
633	B186SDF	Dodge RG13C	Carmichael	WrL	1985	Reserve
652	C832TDF	Dodge RG13TC	Carmichael	WrL	1985	07 Stroud
654	D272BFH	GMC Chevrolet K30 4x4	Telehoist	RIP	1986	Withdrawn
665	D277BFH	GMC Chevrolet K30 4x4	Carmichael/Gloucester FB	RT	1986	Reserve
676	E505JFH	Leyland Freighter 16.17	Fulton & Wylie	WrL	1988	05 Gloucester
677	E506JFH	Leyland Freighter 16.17	Fulton & Wylie	WrL	1988	Reserve
678	E507JFH	Leyland Freighter 16.17	Fulton & Wylie	WrL	1988	Reserve
679	E508JFH	Leyland Freighter 16.17	Fulton & Wylie	WrL	1987	Reserve
685	E624KDF	GMC 3500 4x4	Gloucester FB	RIP	1988	Reserve
686	E368LFH	Iveco-Magirus 140-25A	Carmichael/Magirus DLK23-12	TL	1988	05 Gloucester
687	E625KDF	GMC 3500 4x4	Gloucester FB	RIP	1988	18 Stow-on-the-Wold
688	E626KDF	GMC 3500 4x4	Gloucester FB	RIP	1988	14 Tewksbury
689	E628KDF	GMC 3500 4x4	Gloucester FB	RIP	1988	08 Nailsworth
693	F710RDG	Leyland Freighter T45-180	Fulton & Wylie	WrL	1988	12 Cheltenham
694	F711RDG	Leyland Freighter T45-180	Fulton & Wylie	WrL	1989	Reserve
696	F980UDD	GMC High Sierra 3500 4x4	Telehoist	RIP	1989	10 Dursley
697	F776UDD	GMC High Sierra 3500 4x4	Telehoist	RIP	1989	02 Coleford
698	F777UDD	GMC High Sierra 3500 4x4	Telehoist	RIP	1989	03 Cinderford
699	G529BAD	Renault-Dodge RG13TC	Rosenbauer	WrL	1989	Withdrawn
700	G504XFH	DAF 2500D FAS 6x2	Angloco/Bronto Skylift 28-2Ti	AP	1989	12 Cheltenham
701	F261VAD	GMC 3500 4x4	Telehoist	RT	1989	07 Stroud
702	G505XFH	Leyland Freighter 17.18	Rosenbauer/HIAB	ET	1990	05 Gloucester
703	F509WAD	GMC 3500 4x4	Telehoist	RIP	1989	01 Lydney
704	H736JDG	Leyland Freighter 17.18	Rosenbauer/Angloco	WrL	1991	21 Cirencester
705	H737JDG	Leyland Roadrunner 10.15	Rosenbauer/Angloco	C/WrL	1991	04 Newent
706	H738JDG	Leyland Roadrunner 10.15	Rosenbauer/Angloco	C/WrL	1991	09 Wotton-under-Edge
707	H739JDG	Leyland Roadrunner 10.15	Rosenbauer/Angloco	C/WrL	1991	22 Tetbury
708	K853SDF	Dennis Rapier TF202	John Dennis	WrL	1992	12 Cheltenham
709	K283SDF	Dennis Rapier TF202	John Dennis	WrL	1992	12 Cheltenham
710	K277SDF	Leyland-DAF FA45.160	Locomotors	C/WrL	1992	17 Moreton-in-the-Marsh
711	K276SDF	Leyland-DAF FA45.160	Locomotors	C/WrL	1992	20 Fairford
712	K938UDG	Mercedes-Benz 1120AF	Locomotors/HIAB 071	ET	1993	12 Cheltenham
725	L785WFH	Leyland-DAF FA45.160	Penman	C/WrL	1993	15 Winchcombe
726	L786WFH	Leyland-DAF FA45.160	Penman	C/WrL	1993	03 Cinderford
727	L972YDD	Dennis Rapier TF202	John Dennis	WrL	1993	05 Gloucester
728	L973YDD	Dennis Rapier TF202	John Dennis	WrL	1993	05 Gloucester

Opposite, top: **Gloucestershire took delivery of six new Leyland-DAF appliances during 1995. Shown here is N814MFH, an FA45.180 model bodied by W H Bence in Britol. This appliance is a Foam Support Unit and is operated from Stroud fire station.** *Clive Shearman*
Opposite, bottom: **Displaying German ancestry is Gloucestershire's E368LFH, one of only two appliances of this type in Britain. Based at Gloucester fire station, the Iveco-Magirus 140-25A has a Magirus DLK23-12 ladder and Carmichael bodywork.** *Clive Shearman*

Angloco of Batley, West Yorkshire, supplied three Rosenbauer-bodied Compact Water Tender Ladder appliances to Gloucestershire in July 1991. The trio have Leyland Roadrunner chassis frames though later deliveries of C/WrLs have the Roadrunners successor, the Leyland-DAF FA45 range. One of these, 705 (H737JDG) is based at the retained Newent fire station. *Karl Sillitoe*

746	Pod 1	McDonald Kane/Torton	Control Unit	CU	1989	05 Gloucester
747	Pod 3	McDonald Kane/Torton	Display Unit	DU	1989	Headquarters
748	Pod 4	McDonald Kane/Torton	Damage Control Unit	DCU	1989	07 Stroud
749	Pod 2	McDonald Kane/Torton	Incident Support Unit	ISU	1989	21 Cirencester
751	M34FDD	Leyland-DAF FA45.160	Carmichael International	C/WrL	1994	10 Dursley
752	M35FDD	Leyland-DAF FA45.160	Carmichael International	C/WrL	1994	08 Nailsworth
753	M36FDD	Leyland-DAF FA45.160	Carmichael International	C/WrL	1994	01 Lydney
769	N810MFH	Leyland-DAF FA45.160	Frank Guy/Carmichael Int	C/WrL	1995	14 Tewkesbury
770	N811MFH	Leyland-DAF FA45.160	Frank Guy/Carmichael Int	C/WrL	1995	02 Coleford
771	N812MFH	Leyland-DAF FA45.160	Frank Guy/Carmichael Int	C/WrL	1995	18 Stow-on-the-Wold
772	N813MFH	Leyland-DAF FA45.160	Frank Guy/Carmichael Int	C/WrL	1995	06 Painswick
773	N814MFH	Leyland-DAF FA45.180	WH Bence	SU	1995	07 Stroud
774	N815MFH	Leyland-DAF FA60-210	Massey Tankers	WC/FoT	1995	12 Cheltenham
776	N818MFH	Leyland-DAF FA45.160	WH Bence	GPV	1996	Training Centre
781	N816MFH	Leyland-DAF FA45.160	Excalibur CBK	C/WrL	1996	16 Chipping Campden
782	N817MFH	Leyland-DAF FA45.160	Excalibur CBK	C/WrL	1996	19 Northreach
783	N749SDD	Dennis Sabre S411	John Dennis	WrL	1996	7 Stroud
784	N750SDD	Dennis Sabre S411	John Dennis	WrL	1996	21 Cirencester
785	P	Leyland-DAF FA45.160	Excalibur CBK	C/WrL	1997	
786	P	Leyland-DAF FA45.160	Excalibur CBK	C/WrL	1997	
787	P371VDG	Leyland-DAF FA45.160	W H Bence	PM	1996	Training
	P	Dennis Sabre S411	John Dennis	WrL	1997	
	P	Dennis Sabre S411	John Dennis	WrL	1997	

Notes: 608/10/30/1 were originally WrLs

The Fire Brigade Handbook

GRAMPIAN FIRE BRIGADE

Grampian Fire Brigade, 19 North Anderson Drive, Aberdeen AB9 2TP

HSA590	Dennis F8	Dennis	WrL	1952	Headquarters
LSA9	Dennis F14	Dennis/Metz	TL	1955	Headquarters
TST535T	Bombi	Bombi	ATV	1977	Radio Department
LSO481W	Ford D1617	Angloco	WrL	1980	S85 Balmoral Estate
RRS281X	Dodge G1313	Fulton & Wylie	WrL	1981	S86 Alford
RRS282X	Dodge G1313	Fulton & Wylie	WrL	1981	Reserve
WSE290Y	Dodge G13C	Powell Duffryn Multilift	PM	1981	E77 North Anderson Drive
WSE292Y	Dodge G1313	Fulton & Wylie	WrT	1982	Driving School
WSE293Y	Dodge G1313	Fulton & Wylie	WrL	1982	N39 Gordonstoun School
WSE294Y	Dodge G1313	Fulton & Wylie	WrL	1982	N39 Gordonstoun School
A996DSS	Dodge G1313	Carmichael	WrL	1983	N42 Dufftown
A997DSS	Dodge G1313	Carmichael	WrL	1983	N52 Aberchirder
A998DSS	Dodge G1313	Carmichael	WrL	1983	Retained
A995DSS	Scania P92M	Angloco / Metz DLK30	TL	1984	S96 King Street, Aberdeen
B993KSO	Dodge G13C	Mountain Range	WrL	1984	N49 Cullen
B994KSO	Dodge G13C	Mountain Range	WrL	1984	S91 Aboyne
B995KSO	Dodge G13C	Mountain Range	WrL	1984	N40 Forres
B996KSO	Dodge G13C	Mountain Range	WrT	1984	N40 Forres
B997KSO	Dodge G13C	Carmichael	WrL	1984	Reserve
B998KSO	Dodge G10C	Mountain Range	RT	1984	E77 North Anderson Drive
B999KSO	Dodge G10C	Mountain Range	RT	1984	E70 Fraserburgh
C95RSA	Dodge G13C	Mountain Range	WrL	1985	S87 Strathdon
C96RSA	Dodge G13C	Mountain Range	WrT	1985	S89 Braemar
C97RSA	Dodge G13C	Mountain Range	WrL	1985	N51 Banff
C98RSA	Dodge G13C	Mountain Range	WrL	1985	N45 Keith
C99RSA	Dodge G13C	Mountain Range	WrL	1985	E74 Oldmeldrum
C100RSA	Dodge G13C	Mountain Range	WrL	1985	N47 Fochabers
D358VSA	Dodge G13C	Mountain Range	WrL	1986	N38 Lossiemouth
D359VSA	Dodge G13C	Mountain Range	WrL	1986	N50 Portsoy
D360VSA	Dodge G13C	Mountain Range	WrL	1986	N43 Aberlour
D361VSA	Dodge G13C	Mountain Range	WrL	1986	N46 Rothes
D362VSA	Dodge G10C	Fulton & Wylie	DCU	1986	S96 King Street, Aberdeen
D458WSS	Land Rover 110	Land Rover	L4V	1987	E77 North Anderson Drive
E808ASA	Renault-Dodge G13C	Mountain Range	WrL	1987	N48 Buckie
E809ASA	Renault-Dodge G13C	Mountain Range	WrT	1987	N48 Buckie
E810ASA	Renault-Dodge G13C	Mountain Range	WrL	1987	S94 Laurencekirk
E811ASA	Renault-Dodge G13C	Mountain Range	WrL	1987	S93 Inverbervie
E812ASA	Scania P92M	Saxon Sanbec / Simon SS263	HP	1988	E77 North Anderson Drive
F429GSA	Renault-Dodge G13C	Mountain Range	WrL	1988	E67 Insch
F430GSA	Renault-Dodge G13C	Mountain Range	WrL	1988	E69 Macduff
F431GSA	Renault-Dodge G13C	Powell Duffryn Multilift(1995)	PM	1988	Headquarters
F432GSA	Renault-Dodge G13C	Mountain Range	WrT	1988	N41 Tomintoul
G313MSA	Renault-Dodge G13C	Mountain Range	WrL	1989	E72 Maud
G314MSA	Renault-Dodge G13C	Mountain Range	WrL	1989	S90 Ballater
G315MSA	Renault-Dodge G13C	Mountain Range	WrT	1989	S90 Ballater
G316MSA	Renault-Dodge G13C	Mountain Range	WrL	1989	N53 Huntly
G317MSA	Renault-Dodge G13C	Mountain Range	WrT	1989	N53 Huntly
G318MSA	Renault-Dodge G13C	Powell Duffryn Multilift	PM	1990	N37 Elgin
H286SSA	Scania G93M-210	Mountain Range	WrL	1990	E78 Dyce
H287SSA	Scania G93M-210	Mountain Range	WrL	1990	N37 Elgin
H288SSA	Scania G93M-210	Mountain Range	WrT	1990	N37 Elgin
H289SSA	Scania G93M-210	Mountain Range	WrT	1990	S95 Stonehaven
H290SSA	Scania G93M-210	Mountain Range	WrL	1990	S95 Stonehaven
H291SSA	Scania G93M-210	Powell Duffryn Multilift	PM	1990	S97 Altens, Aberdeen
	Kawasaki Mule 2010	Kawasaki	ATV	1990	N37 Elgin
J445XSO	Scania G93M-210	Reliance Mercury	WrL	1991	S92 Banchory
J446XSO	Scania G93M-210	Reliance Mercury	WrT	1991	S92 Banchory
J447XSO	Scania G93M-210	Reliance Mercury	WrL	1991	E76 Kintore
J448XSO	Scania G93M-210	Reliance Mercury	WrL	1991	E70 Fraserburgh
J449XSO	Scania G93M-210	Reliance Mercury	WrT	1991	E70 Fraserburgh
J997XSS	Land Rover 110 Defender TDi	Land Rover	L4V	1991	N37 Elgin
	Scot Track Glencoe Hydrostatic	Scot Track	ATV	1991	N37 Elgin

Grampian Fire Brigade have not received many new appliances recently, although four are expected this year. Six years ago, the brigade took delivery of its first Scania based pumps and these were fitted with Mountain Range bodywork, H289SSA at Stonehaven fire station. *Malcolm Cook*

J124UDU	Land Rover Discovery	Land Rover	FCV	1992	Headquarters
K281FSO	Scania G93M-210	Emergency One	WrL	1992	E73 Ellon
K282FSO	Scania G93M-210	Emergency One	WrT	1992	E73 Ellon
K283FSO	Scania G93M-210	Emergency One	WrL	1992	Driving School
K284FSO	Scania G93M-210	Emergency One	WrL	1992	E68 Turriff
K285FSO	Scania G93M-210	Emergency One	WrT	1992	E68 Turriff
K286FSO	Scania G93M-250	Angloco / Metz DLK30	TL	1992	E71 Peterhead
L741KRS	Mercedes-Benz Unimog 1550L	Emergency One	L4P	1993	N53 Huntly
L743KRS	Scania G93M-210	Emergency One	WrL	1993	E78 Dyce
L745KRS	Scania G93M-210	Emergency One	WrL	1993	E75 Inverurie
L746KRS	Scania G93M-210	Emergency One	WrT	1993	E75 Inverurie
L748KRS	Scania G93M-210	Emergency One	WrL	1993	E71 Peterhead
L749KRS	Scania G93M-210	Emergency One	WrT	1993	E71 Peterhead
L244LRS	Kawasaki Mule	Kawasaki	ATV	1993	E77 North Anderson Drive
M990PSS	Scania G93M-220	Emergency One	WrL	1994	S96 King Street, Aberdeen
M991PSS	Scania G93M-220	Emergency One	WrL	1994	S96 King Street, Aberdeen
M992PSS	Scania G93M-220	Emergency One	WrL	1994	E77 North Anderson Drive
M993PSS	Scania G93M-220	Emergency One	WrL	1994	E77 North Anderson Drive
M994PSS	Scania G93M-220	Emergency One	WrL	1994	S97 Altens, Aberdeen
N402XRS	Scania P93HKZ-250	Powell Duffryn Multilift/HIAB	PM	1996	E71 Peterhead
Pod 1	Kingswells	Flat Bed Lorry	FBL	1982	Workshops
Pod 2	Kingswells	Canteen	CaV	1982	E77 North Anderson Drive
Pod 3	Kingswells	Chemical Incident Unit	CIU	1993	S97 Altens, Aberdeen
Pod 4	Kingswells	B.A.Training Unit	BATU	1993	S97 Altens, Aberdeen
Pod 5	Kingswells	Hi-Expansion Foam Unit	HiEx	1984	E77 North Anderson Drive
Pod 6	Kingswells	Contol Unit	CU	1984	E77 North Anderson Drive
Pod 8	Kingswells	5000 Litre Foam Bowser	FoT	1985	S97 Altens, Aberdeen
Pod 9	Kingswells	Display Unit	DisU	1988	E77 North Anderson Drive
Pod 10	Kingswells	Helicopter Support Unit	HelU	1990	N37 Elgin
Pod 11	Kingswells	Hose Layer	HL	1991	Workshops
Pod 12	Emergency One	Heavy Rescue Unit	HRU	1996	E71 Peterhead

Note: F431GSA was new as a WrL.

GREATER MANCHESTER COUNTY FIRE SERVICE

Greater Manchester County Fire Service, 146 Bolton Road, Pendlebury,
Manchester M27 8US

PDK717	Dennis F21	Dennis/Metz	TL	1957	Rochdale FS Museum
Trailer	Entwhistle	Fuel Tanker		1977	Workshops
Boat 1	Zodiac	Rescue Boat & trailer	FBt	1977	C32 Heywood
TBA260R	Ford D1114	Angloco	BAT	1977	E56 Mossley (R)
GRJ298V	Ford D1617	Cheshire Fire Engineering	FoT	1980	E52 Philips Park
LNE536V	Ford D1617	Cheshire Fire Engineering	FoT	1980	D47 Cheadle Hulme
PNB480W	Dodge G1811P	Artic Towing Unit	TU	1981	A12 Agecroft
BVM543Y	Shelvoke & Drewry WY	Angloco/Simon Snorkel SS263	HP	1983	A15 Stretford (R)
BVM544Y	Shelvoke & Drewry WY	Angloco/Simon Snorkel SS263	HP	1983	D44 Moss Side (R)
FVR55Y	Land Rover 110	Mountain Range	LRV	1983	C32 Heywood
FVR56Y	Land Rover 110	Mountain Range	LRV	1983	C37 Whitefield(R)
A24JDB	Dennis RS131	Dennis/GMCFS	WrL	1983	Publicity - Welephant
A649OJA	Dodge G13	Mountain-Range/GMCFS	WrL	1984	Publicity - Welephant
A651OJA	Dodge G13	Mountain-Range/GMCFS	WrL	1984	Publicity - Welephant
A959ONB	Dodge G16C	Carmichael	FoT	1984	A16 Sale
B967RBA	Dodge G13 / Astatic Skidmaster	Mountain-Range	WrL	1984	Driving School
C695AJA	Bedford CF2-350/FF 4x4	Fulton & Wylie	L4P	1985	Cadets
C434AND	Renault-Dodge G16C	Saxon Sanbec/Simon SS263	HP	1985	B20 Bolton
C42YBA	Renault-Dodge G11C	Fulton & Wylie	HL	1985	B23 Farnworth
C747DNF	Mountain Range	Mountain Range	WrL	1986	Publicity - Welephant
C748DNF	Renault-Dodge G13	Mountain Range	WrL	1986	Publicity - Welephant
C750DNF	Reanult-Dodge G13	Mountain Range	WrL	1986	Cadets
C752DNF	Renault-Dodge G13	Mountain Range	WrL	1986	Cadets
C538YRJ	Leyland Roadrunner	GMCFS	HFU	1986	B23 Farnworth
D765NNB	Renault-Dodge G13C	Fulton & Wylie	WrL	1987	E53 Gorton (R)
E854UNA	Leyland Freighter T180	Fulton & Wylie	WrL	1987	Driving School
E777WBU	Renault-Dodge G16	Saxon Sanbec / Simon SS263	HP	1988	E50 Manchester Central
E192WNA	Renault-Dodge G13C	Fulton & Wylie	WrL	1987	E52 Phillips Park, Manchester
E194WNA	Renault-Dodge G13C	Fulton & Wylie	WrL	1987	A17 Altrincham (R)
E195WNA	Renault-Dodge G13C	Fulton & Wylie	WrL	1987	B24 Wigan (R)
E196WNA	Renault-Dodge G13C	Fulton & Wylie	WrL	1988	Driving School
E198WNA	Renault-Dodge G13C	Fulton & Wylie	WrL	1988	D41 Stockport (R)
E457WND	Dennis SS137	Fulton & Wylie	WrL	1988	A15 Stretford (R)
E458WND	Dennis SS137	Fulton & Wylie	WrL	1988	Workshops (R)
E459WND	Dennis SS137	Fulton & Wylie	WrL	1988	E55 Stalybridge (R)
E460WND	Dennis SS137	Fulton & Wylie	WrL	1988	Workshops (R)
E461WND	Dennis SS137	Fulton & Wylie	WrL	1988	Workshops (R)
E462WND	Dennis SS137	Fulton & Wylie	WrL	1988	Workshops (R)
F674GNC	Renault-Dodge G16	Saxon Sanbec / Simon SS263	HP	1989	D44 Moss Side
F62ENB	Renault-Dodge G13	Mountain Range	WrL	1988	C36 Bury (R)
F63ENB	Renault-Dodge G13	Mountain Range	WrL	1988	D40 Whitehill, Stockport (R)
F64ENB	Renault-Dodge G13	Mountain Range	WrL	1988	B20 Bolton (R)
F65ENB	Renault-Dodge G13	Mountain Range	WrL	1988	C35 Chadderton (R)
F66ENB	Renault-Dodge G13	Mountain Range	WrL	1988	C34 Hollins (R)
F67ENB	Renault-Dodge G13	Mountain Range	WrL	1989	Training Centre
F68ENB	Renault-Dodge G13	Mountain Range	WrL	1989	Training Centre
F37HND	Renault-Dodge G13TC	Mountain Range	WrL	1989	Driving School
F38HND	Renault-Dodge G13TC	Mountain Range	WrL	1989	Driving School
F39HND	Renault-Dodge G13TC	Mountain Range	WrL	1989	Driving School
F41HND	Renault-Dodge G13TC	Mountain Range	WrL	1989	E56 Mossley
F42HND	Renault-Dodge G13TC	Mountain Range	WrL	1989	D43 Marple
G103RBA	Renault-Dodge G13TC	Fulton & Wylie	WrL	1989	C38 Ramsbottom
G104RBA	Renault-Dodge G13TC	Fulton & Wylie	WrL	1989	C38 Ramsbottom
G105RBA	Renault-Dodge G13TC	Fulton & Wylie	WrL	1989	D41 Stockport
G106RBA	Renault-Dodge G13TC	Fulton & Wylie	WrL	1989	D41 Stockport
G107RBA	Renault-Dodge G13TC	Fulton & Wylie	WrL	1989	Driving School
G108RBA	Renault-Dodge G13TC	Fulton & Wylie	WrL	1989	Driving School

The last batches of Renault-Dodge G13 Water Tender Ladder appliances with Greater Manchester are now filtering into the Reserve and Training school fleets. Mountain Range-bodied F62ENB from 1988 was on the run at Altrincham where it was photographed in the town centre though it, too, has since moved to be reserve at Bury. *Tony Wilson*

G412RJA	Leyland Freighter T45-180	Fulton & Wylie	WrL	1990	A16 Sale
G413RJA	Leyland Freighter T45-180	Fulton & Wylie	WrL	1990	A16 Sale
G414RJA	Leyland Freighter T45-180	Fulton & Wylie	WrL	1990	B20 Bolton
G415RJA	Leyland Freighter T45-180	Fulton & Wylie	WrL	1990	B20 Bolton
G416RJA	Leyland Freighter T45-180	Fulton & Wylie	WrL	1990	B25 Hindley
G417RJA	Leyland Freighter T45-180	Fulton & Wylie	WrL	1990	B26 Atherton
G418RJA	Leyland Freighter T45-180	Fulton & Wylie	WrL	1990	E51 Blackley
G419RJA	Leyland Freighter T45-180	Fulton & Wylie	WrL	1990	E51 Blackley
G199SRJ	Renault-Dodge G16	Saxon Sanbec / Simon SS263	HP	1990	C33 Oldham
H244CVR	Renault-Dodge S56	Harrop / GMCFS	CU	1990	A12 Agecroft
H501DND	Dennis F127	Saxon Sanbec / Simon SS263	HP	1990	A15 Stretford
H374FBA	Leyland Freighter T45-210	Fulton & Wylie	WrL	1991	A11 Broughton
H375FBA	Leyland Freighter T45-210	Fulton & Wylie	WrL	1991	A11 Broughton
H376FBA	Leyland Freighter T45-210	Fulton & Wylie	WrL	1991	A10 Salford
H377FBA	Leyland Freighter T45-210	Fulton & Wylie	WrL	1991	A10 Salford
H378FBA	Leyland Freighter T45-210	Fulton & Wylie	WrL	1991	C31 Littleborough
H379FBA	Leyland Freighter T45-210	Fulton & Wylie	WrL	1991	B22 Horwich
H380FBA	Leyland Freighter T45-210	Fulton & Wylie	WrL	1991	C31 Littleborough
H381FBA	Leyland Freighter T45-210	Fulton & Wylie	WrL	1991	B22 Horwich
H382FBA	Leyland-DAF 60-210	Fulton & Wylie	WrL	1991	C34 Hollins
H383FBA	Leyland Freighter T45-210	Fulton & Wylie	WrL	1991	D44 Moss Side
H384FBA	Leyland Freighter T45-210	Fulton & Wylie	WrL	1991	D44 Moss Side
H385FBA	Leyland Freighter T45-210	Fulton & Wylie	WrL	1991	E50 Manchester Central
H386FBA	Leyland Freighter T45-210	Fulton & Wylie	WrL	1991	E50 Manchester Central
J250PJA	Leyland-DAF Roadrunner 10.15	Whiteacres / GMCFS	EST	1992	C30 Rochdale (R)
J561PND	Volvo FL10	Bedwas / Simon Snorkel ST290S	ALP	1992	A10 Salford (R)
Boat 2	Zodiac	Rescue Boat & Trailer	IRBt	1992	A13 Eccles
K203VVU	Leyland-DAF 60-210Ti	Reliance Mercury	WrL	1992	A13 Eccles
K204VVU	Leyland-DAF 60-210Ti	Reliance Mercury	WrL	1992	A13 Eccles
K205VVU	Leyland-DAF 60-210Ti	Reliance Mercury	WrL	1992	B23 Farnworth
K206VVU	Leyland-DAF 60-210Ti	Reliance Mercury	WrL	1992	B23 Farnworth
K207VVU	Leyland-DAF 60-210Ti	Reliance Mercury	WrL	1992	C37 Whitefield
K208VVU	Leyland-DAF 60-210Ti	Reliance Mercury	WrL	1992	A15 Stretford

Greater Manchester County Fire Service normally allocate matching pairs of pumping appliances to all two-pump stations. Various permutations of chassis and body manufacturers have featured in orders placed in the 1990s. M697NNC a Volvo FL6.14/Saxon Sanbec, provides fire cover for Oldham with sister M696NNC. *Gavin Stewart*

K209VVU	Leyland-DAF 60-210Ti	Reliance Mercury	WrL	1992	A15 Stretford
K210VVU	Leyland-DAF 60-210Ti	Reliance Mercury	WrL	1992	A12 Agecroft
K211VVU	Leyland-DAF 60-210Ti	Reliance Mercury	WrL	1993	A14 Irlam
K212VVU	Leyland-DAF 60-210Ti	Reliance Mercury	WrL	1993	A14 Irlam
K681XBU	Peugeot-Talbot Express	Talbot	ART	1993	Training Centre
K128XRJ	Leyland-DAF Roadrunner 45-160	Bedwas	BAT	1993	B21 Hall'i'th'wood, Bolton North
K129XRJ	Leyland-DAF Roadrunner 45-160	Bedwas	BAT	1993	E57 Hyde
K64XBA	Volvo FL10	Bedwas / Simon Snorkel ST290S	ALP	1993	B27 Leigh
K65XBA	Volvo FL10	Bedwas / Simon Snorkel ST290S	ALP	1993	D40 Whitehill, Stockport
K783ANC	Renault Midliner M230-15D	HCB-Angus	WrL	1993	E54 Ashton under Lyne
K784ANC	Renault Midliner M230-15D	HCB-Angus	WrL	1993	E54 Ashton under Lyne
K785ANC	Renault Midliner M230-15D	HCB-Angus	WrL	1993	C30 Rochdale
K786ANC	Renault Midliner M230-15D	HCB-Angus	WrL	1993	C30 Rochdale
K787ANC	Renault Midliner M230-15D	HCB-Angus	WrL	1993	C36 Bury
K788ANC	Renault Midliner M230-15D	HCB-Angus	WrL	1993	C36 Bury
K789ANC	Renault Midliner M230-15D	HCB-Angus	WrL	1993	D40 Whitehill, Stockport
L609DDB	Dennis DFS237	Saxon Sanbec	WrL	1993	B21 Hall'i'th'wood, Bolton North
L610DDB	Dennis DFS237	Saxon Sanbec	WrL	1993	B24 Wigan
L611DDB	Dennis DFS237	Saxon Sanbec	WrL	1993	B24 Wigan
L612DDB	Dennis DFS237	Saxon Sanbec	WrL	1993	C32 Heywood
L613DDB	Dennis DFS237	Saxon Sanbec	WrL	1993	C32 Heywood
L614DDB	Dennis DFS237	Saxon Sanbec	WrL	1993	E57 Hyde
L615DDB	Dennis DFS237	Saxon Sanbec	WrL	1993	D46 Wythenshaw
L616DDB	Dennis DFS237	Saxon Sanbec	WrL	1993	D46 Wythenshaw
L617DDB	Dennis DFS237	Saxon Sanbec	WrL	1993	E55 Stalybridge
L618DDB	Dennis DFS237	Saxon Sanbec	WrL	1993	E55 Stalybridge
Trailer 4030	Lambirding	Lighting unit	LU	1994	C38 Ramsbottom
Trailer 4031	Lambirding	Lighting unit	LU	1994	B21 Hall'i'th'wood, Bolton North
Trailer 4032	Lambirding	Lighting unit	LU	1994	D43 Marple
L978ENC	Leyland-DAF Roadrunner 45-160	Pickering/Bedwas	EST	1994	A13 Eccles
L979ENC	Leyland-DAF Roadrunner 45-160	Pickering/Bedwas	EST	1994	E53 Gorton
L980ENC	Leyland-DAF Roadrunner 45-160	Pickering/Bedwas	EST	1994	D42 Offerton
L981ENC	Leyland-DAF Roadrunner 45-160	Pickering/Bedwas	EST	1994	B24 Wigan

All five of Greater Manchester's front line Emergency Salvage Tenders now have Leyland-DAF 45.160Ti chassis. M364SDB, show here in Rochdale, dates from 1995. *Gavin Stewart*

M317NVU	Peugeot-Talbot Express	Talbot/Bedwas	LFA	1994	A13 Eccles
M318NVU	Peugeot-Talbot Express	Talbot/Bedwas	LFA	1994	B21 Hall'i'th'wood, Bolton North
M319NVU	Peugeot-Talbot Express	Talbot/Bedwas	LFA	1994	C38 Ramsbottom
M320NVU	Peugeot-Talbot Express	Talbot/Bedwas	LFA	1994	D43 Marple
M321NVU	Peugeot-Talbot Express	Talbot/Bedwas	LFA	1994	E56 Mossley
M691NNC	Volvo FL6-14	Saxon Sanbec	WrL	1995	B27 Leigh
M692NNC	Volvo FL6-14	Saxon Sanbec	WrL	1995	B27 Leigh
M693NNC	Volvo FL6-14	Saxon Sanbec	WrL	1995	D45 Withington
M694NNC	Volvo FL6-14	Saxon Sanbec	WrL	1995	E52 Philips Park, Manchester
M695NNC	Volvo FL6-14	Saxon Sanbec	WrL	1995	E52 Philips Park, Manchester
M696NNC	Volvo FL6-14	Saxon Sanbec	WrL	1995	C33 Oldham
M697NNC	Volvo FL6-14	Saxon Sanbec	WrL	1995	C33 Oldham
M364SDB	Leyland-DAF 45-160	Whiteacre/Bedwas	EST	1994	C30 Rochdale
N665WVR	Ford Transit VE6	Ford/Bedwas	CaV	1995	B20 Bolton
	Moffett Mounty M2403N	*Normally carried by C42YBA*	FLT	1995	B23 Farnworth
N374YNC	Volvo FL10	Saxon Sanbec/Simon ST290-S	ALP	1996	A10 Salford
N375YNC	Volvo FL10	Saxon Sanbec/Simon ST290-S	ALP	1996	C36 Bury
N376YNC	Volvo FL10	Saxon Sanbec/Simon ST290-S	ALP	1996	E50 Manchester Central
N377YNC	Volvo FL10	Saxon Sanbec/Simon ST290-S	ALP	1996	E54 Ashton-u-Lyne
N651YNC	Volvo FL6-14	Saxon Sanbec	WrL	1996	A17 Altrincham
N652YNC	Volvo FL6-14	Saxon Sanbec	WrL	1996	D42 Offerton
N653YNC	Volvo FL6-14	Saxon Sanbec	WrL	1996	D47 Cheadle Hulme
N654YNC	Volvo FL6-14	Saxon Sanbec	WrL	1996	C35 Chadderton
N655YNC	Volvo FL6-14	Saxon Sanbec	WrL	1996	C35 Chadderton
N656YNC	Volvo FL6-14	Saxon Sanbec	WrL	1996	B21Hall'ith'Wood, Bolton North
N657YNC	Volvo FL6-14	Saxon Sanbec	WrL	1996	E53 Gorton
N635XBA	ERF EC8	Clayton Tankers/Saxon Sanbec	FoT	1996	E52 Philips Park, Manchester
P330FVR	ERF EC8	Clayton Tankers/Saxon Sanbec	FoT	1996	
P331FVR	ERF EC8	Clayton Tankers/Saxon Sanbec	FoT	1997	
P332FVR	ERF EC8	Clayton Tankers/Saxon Sanbec	FoT	1997	
P	ERF EC8	Don Bur	SIU	1996	B23 Farnworth
P21GNA	Volvo B6-9.9m	Northern Counties/Saxon	ICU	1996	A12 Agecroft

HAMPSHIRE FIRE & RESCUE SERVICE

Hampshire Fire & Rescue Service, Leigh Road, Eastleigh, Hampshire SO5 4SJ

GOT356K	Leyland Mastiff	Fergusson	WrC	1972	B17 Fareham
GOT357K	Leyland Mastiff	Fergusson	WrC	1972	D47 Fordingbridge
NHO554L	Bedford TKG	Hampshire FB	CaV	1973	C30 Winchester
GPX583N	Land Rover Series III 109	Hampshire FB	L4P	1975	D47 Fordingbridge
GPX584N	Land Rover Series III 109	Hampshire FB	L4T	1975	B16 Havant
GPX586N	Land Rover Series III 109	Hampshire FB	L4P	1975	D55 Hamble
KPX236P	Land Rover Series III 109	Hampshire FB	L4T	1976	C36 Alresford
KPX237P	Land Rover Series III 109	Hampshire FB	L4P	1976	C32 Eastleigh
KPX238P	Land Rover Series III 109	Hampshire FB	L4P	1976	A05 Alton
KPX239P	Land Rover Series III 109	Hampshire FB	L4T	1976	D52 Burley
KPX240P	Land Rover Series III 109	Hampshire FB	L4T	1976	A04 Fleet
KTR891P	Leyland Mastiff	Fergusson/HFB	WrC	1976	D58 Hardley
LOW465R	Dodge K1613	Hampshire FB/Wreckers(1984)	RV	1977	Workshops
MBK390R	Bedford VAS5	Hampshire FB	CU	1977	C30 Winchester
OOW53S	Bedford TKG	HCB-Angus	WrT	1977	C Reserve
OOW55S	Bedford TKG	HCB-Angus CSV	WrT	1977	Fire Safety
OOW57S	Bedford TKG	HCB-Angus CSV	WrL	1978	A Reserve
PPX511S	Land Rover Series III 109	Land Rover/HFB	L4T	1978	A08 Hartley Wintney
PPX512S	Land Rover Series III 109	Land Rover/HFB	L4V	1978	Driving School
PPX513S	Leyland Mastiff	Fergusson	WC/FoT	1978	C31 Andover
PPX519S	Bedford TKG	HCB-Angus CSV	WrL	1978	Unallocated
ROT687S	Land Rover Series III 109	Hampshire FB	L4T	1978	B17 Fareham
ROT688S	Land Rover Series III 109	Hampshire FB	L4P	1978	D58 Hardley
ROT689S	Land Rover Series III 109	Hampshire FB	L4T	1978	D44 Hythe
TRV403T	Bedford TKG	HCB-Angus CSV	WrT	1979	D Reserve
TRV404T	Bedford TKG	HCB-Angus CSV	WrT	1979	C41 Droxford
TRV405T	Bedford TKG	HCB-Angus CSV	WrL	1979	C Reserve
XOT981V	Dennis RS133	Hampshire FB	WrT	1980	A Reserve
XOT982V	Dennis RS133	Hampshire FB	WrL	1980	D Reserve
XOT983V	Dennis RS133	Hampshire FB	WrT	1980	Unallocated
XOT984V	Dennis RS133	Hampshire FB	WrT	1980	B Reserve
XOT985V	Bedford TKG	HCB-Angus CSV	WrL	1980	B21 Hayling Island

Photographed when passing through Basingstoke was G162UPO. The VolvoFL6 carries a body built by Hampshire FB for use as a Special Equipment Unit.
Gerald Mead

Reg	Chassis	Body	Type	Year	Station
BTP483W	Bedford TKG	HCB-Angus CSV	WrL	1981	A Reserve
BTP489W	Dennis RS133	Hampshire FB	WrT	1981	Unallocated
BTP490W	Dennis RS133	Hampshire FB	WrL	1981	Training Centre
BTP491W	Bedford TKG	HCB-Angus HSC	WrL	1981	B Reserve
FPO601X	Bedford TKG	Hampshire FB	WrT	1981	B27 Titchfield
FPO602X	Bedford TKG	Hampshire FB	WrT	1981	Training Centre
FPO603X	Bedford TKG	HCB-Angus HSC	WrT	1982	B28 Portchester
FPO604X	Bedford TKG	HCB-Angus HSC	WrT	1982	A Reserve
FPO605X	Bedford TKG	HCB-Angus HSC	WrL	1982	B Reserve
FPO608X	Land Rover Series III 109	Hampshire FB	L4T	1982	A01 Basingstoke
DPR212Y	Bedford TKG	HCB-Angus HSC/Hampshire FB	WrT	1983	Driving School
JTP634Y	Bedford TKG	Hampshire FB	WrT	1982	C Reserve
JTP638Y	Bedford TKG	Hampshire FB	WrT	1983	Training Centre
A32OPX	Bedford TKG	HCB-Angus HSC	WrT	1983	C Reserve
A33OPX	Bedford TKG	HCB-Angus HSC	WrT	1983	B22 Wickham
A34OPX	Bedford TKG	HCB-Angus HSC	WrL	1983	Training Centre
A35OPX	Bedford TKG	HCB-Angus HSC	WrT	1983	A13 Liphook
A36OPX	Dennis RS133	Hampshire FB	WrT	1983	A01 Basingstoke
A336RBK	Land Rover Series III 109	Hampshire FB	L4T	1984	D48 Lyndhurst
A337RBK	Land Rover Series III 109	Hampshire FB	L4T	1984	A13 Liphook
A338RBK	Dennis RS133	Hampshire FB	WrT	1984	D Reserve
B163TPX	Land Rover Series III 109	Hampshire FB	L4T	1984	A02 Rushmore, Farnborough
B164TPX	Bedford TKG	Hampshire FB	WrT	1984	C33 Romsey
B165TPX	Bedford TKG	HCB-Angus HSC	WrT	1985	B19 Waterlooville
B166TPX	Bedford TKG	HCB-Angus HSC	WrT	1985	A04 Fleet
B167TPX	Bedford TKG	Hampshire FB	WrT	1985	D55 Hamble
B168TPX	Bedford TKG	Hampshire FB	WrT	1985	Workshops
Boat	Watercraft R5	Inflatable boat & trailer	IRBt	1985	D55 Hamble
C844YCR	Bedford TKG	Hampshire FB	WrT	1985	A05 Alton
C847YCR	Dennis RS135	Hampshire FB	WrT	1986	Driving School
C848YCR	Bedford TKG	HCB-Angus HSC	WrT	1986	D45 Ringwood
C849YCR	Bedford TKG	HCB-Angus HSC	WrT	1986	B29 Petersfield
D373DBK	Bedford TKG	HCB-Angus HSC	WrT	1986	D51 New Milton
D374DBK	Dennis RS135	Hampshire FB	WrT	1986	A12 Tadley
D375DBK	Dennis RS135	Hampshire FB	WrT	1986	B Reserve
D622DTR	Dennis RS135	Hampshire FB	WrL	1986	D Reserve
D623DTR	Renault-Dodge G16	Saxon/Simon Snorkel SS70	HP	1987	D53 Redbridge, Southampton
D624DTR	Leyland Freighter 16.17	Buckingham Tankers/HFB	WrC	1987	01 Basingstoke
D626DTR	Bedford TL	HCB-Angus HSC	WrT	1987	D43 Lymington
E741HRV	Renault-Dodge G16	Saxon/Simon Snorkel SS70	HP	1987	Unallocated
E748HRV	Bedford TKG	Hampshire FB HSC	WrL	1987	A07 Grayshott
E750HRV	Dennis DS153	HCB-Angus	WrT	1987	D50 Brockenhurst
E751HRV	Dennis DS153	HCB-Angus	WrT	1987	B21 Hayling Island
E752HRV	Dennis DS153	HCB-Angus	WrT	1987	D52 Burley
E753HRV	Dennis DS153	HCB-Angus	WrT	1987	A09 Kingsclere
E754HRV	Bedford TL	Hampshire FB HSC	WrL	1988	D43 Lymington
E755HRV	Bedford TL	Hampshire FB HSC	WrT	1988	D49 Beaulieu
F986NRV	Volvo FL6.14	Metz/Metz DL30	TL	1988	B24 Southsea, Portsmouth
F987NRV	Volvo FL6.14	HCB-Angus	WrT	1988	A02 Rushmore, Farnborough
F988NRV	Volvo FL6.14	HCB-Angus	WrT	1988	A10 Odiham
F989NRV	Volvo FL6.14	Hampshire FB	WrL	1989	C40 Bishops Waltham
F990NRV	Volvo FL6.14	Hampshire FB	WrL	1988	A06 Whitchurch
F991NRV	Volvo FL6.14	HCB-Angus	WrL	1989	Training Centre
F992NRV	Volvo FL6.14	HCB-Angus	WrL	1989	A08 Hartley Wintney
F994NRV	Renault-Dodge G17	Saxon/Simon Snorkel SS70	HP	1988	A01 Basingstoke
F31NTP	Land Rover 110	Hampshire FB	L4T	1989	A03 Bordon
G162UPO	Volvo FL6.14	Hampshire FB	SEU	1989	A01 Basingstoke
G163UPO	Volvo FL6.14	Hampshire FB	SEU	1990	D54 St Mary's, Southampton
G164UPO	Volvo FL6.14	HCB-Angus	WrT	1990	C38 Botley
G165UPO	Volvo FL6.14	HCB-Angus	WrT	1990	A14 Yateley
G166UPO	Volvo FL6.14	HCB-Angus	WrT	1990	A03 Bordon
G167UPO	Volvo FL6.14	HCB-Angus	WrT	1990	B25 Horndean
G168UPO	Volvo FL6.14	HCB-Angus	WrL	1990	D58 Hardley
G169UPO	Volvo FL6.14	HCB-Angus	WrL	1990	D45 Ringwood
G179UPO	Volvo FL6.14	Saxon Sanbec	WrL	1990	C36 Alresford
G180UPO	Volvo FL6.14	Saxon Sanbec	WrT	1990	Unallocated
G181UPO	Volvo FL6.14	Saxon Sanbec	WrL	1990	C33 Romsey
G186UPO	Renault-Dodge G17	Angloco / Simon Snorkel SS70	HP	1990	Unallocated
H370BTP	Dennis DS153	Saxon Sanbec	WrT	1990	B26 Emsworth
H371BTP	Volvo FL6.14	Saxon Sanbec	WrL	1990	D47 Fordingbridge

The Fire Brigade Handbook

H372BTP	Volvo FL6.14	Saxon Sanbec	WrL/R	1990	B29 Petersfield
H373BTP	Volvo FL6.14	Saxon Sanbec	WrL	1990	Unallocated
H374BTP	Volvo FL6.14	HCB-Angus	WrL	1991	D51 New Milton
H375BTP	Volvo FL6.14	HCB-Angus	WrT	1991	C35 Sutton Scotney
H376BTP	Volvo FL6.14	HCB-Angus	WrL	1991	A05 Alton
J101HBP	Volvo FL6.14	HCB-Angus	WrT	1992	A11 Overton
J102HBP	Volvo FL6.14	HCB-Angus	WrL/R	1992	A02 Rushmoor, Farnborough
J103HBP	Volvo FL6.14	HCB-Angus	WrL/R	1992	C31 Andover
J104HBP	Volvo FL6.14	HCB-Angus	WrL/R	1992	Unallocated
J105HBP	Volvo FL6.14	HCB-Angus	WrL/R	1992	Unallocated
J115HBP	Volvo FL6.14	HCB-Angus	WrT	1992	C37 Twyford
K182MPO	Volvo FL6.14	Hampshire FRS	SEU	1992	B20 Copnor Rd, Portsmouth
K183MPO	Volvo FL6.14	HCB-Angus	WrT	1993	D46 Totton
K184MPO	Volvo FL6.14	HCB-Angus	WrL	1993	B18 Gosport
K185MPO	Volvo FL6.14	HCB-Angus	WrL	1993	Unallocated
K186MPO	Volvo FL6.14	HCB-Angus	WrL	1992	B19 Waterlooville
K187MPO	Volvo FL6.14	HCB-Angus	WrL	1993	C34 Stockbridge
K188MPO	Volvo FL6.14	HCB-Angus	WrT	1993	Unallocated
K189MPO	Volvo FL6.14	HCB-Angus	WrT	1993	D44 Hythe
K192MPO	Volvo FL6.14	HCB-Angus	WrT	1993	C32 Eastleigh
K198MPO	Volvo FL6.14	HCB-Angus	WrT	1993	B16 Havant
K199MPO	Iveco-Ford Cargo 110E15	Hampshire FRS	GPV	1993	Driving School
K355MPO	Iveco-Ford Cargo 110E15	Ray Smith	PM	1993	Driving School
Pod	Ray Smith/Musslewhite	Flat bed lorry	FBL	1993	Driving School
Pod	Ray Smith/Musslewhite	General purpose lorry	GPV	1993	Driving School
Pod	Ray Smith/Bence	Exhibition Unit	ExU	1993	Headquarters
L79RTP	Volvo FL6.14	HCB-Angus	WrL	1994	B24 Southsea, Porthsmouth
L81RTP	Volvo FL6.18	Metz/Metz DLK30	TL	1994	C30 Winchester
L84RTP	Volvo FL6.14	HCB-Angus	WrL/R	1994	D48 Lyndhurst
L85RTP	Volvo FL6.14	HCB-Angus	WrT	1994	D54 St Mary's, Southampton
L87RTP	Volvo FL6.14	HCB-Angus	WrL	1994	B17 Fareham
L89RTP	Volvo FL6.14	HCB-Angus	WrT	1994	B23 Cosham, Portsmouth
M267XOT	Volvo FL7	Locomotors/HFRS	WC	1995	Unallocated
M268XOT	Iveco-Ford Cargo 120E15	Ray Smith	PM	1994	Driving School
M270XOT	Volvo FL6.14	John Dennis	WrT	1995	A01 Basingstoke
M272XOT	Land Rover Defender 110	Land Rover/Hampshire FRS	L4T	1995	D45 Ringwood
M273XOT	Volvo FL6.14	John Dennis	WrT	1995	C31 Andover
M274XOT	Volvo FL6.14	John Dennis	WrT	1995	B18 Gosport
M276XOT	Iveco-Ford Cargo 120E15	Ray Smith	GPV	1994	Driving School
N104EBP	Volvo FL6.14	John Dennis	WrL/R	1995	Driving School
N105EBP	Volvo FL6.14	John Dennis	WrL	1995	D53 Redbridge, Southampton
N106EBP	Volvo FL6.14	John Dennis	WrL	1996	B23 Cosham, Portsmouth
N107EBP	Volvo FL6.14	John Dennis	WrL	1996	B20 Copnor Rd, Portsmouth
N108EBP	Volvo FL6.14	John Dennis	WrL	1996	A01 Basingstoke
N109EBP	Iveco-Ford Cargo 135E18	Angloco/HIAB 071	MRV	1995	D48 Lyndhurst
N117EBP	Volvo FL7	Angloco	WC/FoT	1995	C32 Eastleigh
N121EBP	Volvo FL6.14	Saxon Sanbec	WrT	1996	D56 Hightown, Southampton
N122EBP	Volvo FL6.14	Saxon Sanbec	WrL	1996	D56 Hightown, Southampton
N123EBP	Volvo FL6.14	Saxon Sanbec	WrL	1996	B16 Havant
N124EBP	Volvo FL6.14	Saxon Sanbec	WrT	1996	D53 Redbridge, Southampton
N325HBK	Volvo FL6.14	Saxon Sanbec	WrT	1996	C30 Winchester
N325HBK	Volvo FL6.14	Saxon Sanbec	WrT	1996	A02 Rushmoor, Farnborough
P912KPX	Iveco-Ford Cargo 135E18	Angloco/HIAB 071	ATV	1996	A02 Rushmoor, Farnborough
P	Iveco-Ford Cargo 135E18	Angloco/HIAB 071	ATV	1997	
P950JTR	Volvo FL6.14	Saxon Sanbec	WrT	1996	C32 Eastleigh
P951JTR	Volvo FL6.14	Saxon Sanbec	WrT	1996	B24 Southsea
P952JTR	Volvo FL6.14	Saxon Sanbec	WrL	1996	D54 St Marys, Southampton
P953JTR	Volvo FL6.14	Saxon Sanbec	WrL/R	1996	C32 Eastleigh
P960JTR	Volvo FL6.14	Saxon Sanbec	WrL/R	1996	C30 Winchester
P964JTR	Volvo FL6.14	Saxon Sanbec	WrL/R	1996	B17 Fareham
P	Volvo FL6.14	Saxon Sanbec	WrL	1996	
P	Volvo FL6.14	Saxon Sanbec	WrL	1996	
P	Volvo FL6.14	Saxon Sanbec	WrL	1996	
P	Volvo FL6.14	Saxon Sanbec	WrL	1996	
P	Volvo FL6.14	Saxon Sanbec	WrL	1996	
P	Volvo FL6.14	Saxon Sanbec	WrL	1996	

KOW286P was previously BAT; LOW465R was previously a Simonitor

HEREFORD & WORCESTER FIRE BRIGADE

Hereford & Worcester Fire Brigade, Copenhagen Street, Worcester WR1 2HQ

HUY878T	Land Rover Series III 88	Land Rover/H&WFB	L4V	1978	21 Worcester
RUY150V	Dennis F124	Dennis / Simon Snorkel SS263	HP	1980	24 Kidderminster
YAB257W	Dodge G13	Carmichael Benson	CU	1980	41 Malvern
HAB900X	Dennis DF133	Dennis	WrC	1981	46 Hereford
ONP742Y	Dennis RS133	Dennis	WrL	1982	Training Centre
ONP743Y	Dennis RS131	Dennis	WrL	1982	Training Centre
A528WNP	Land Rover Series III 6x6	Carmichael	L6P	1983	22 Stourport-on-Severn
A801AWP	Dennis RS133	Dennis	WrL	1984	Reserve
A802AWP	Dennis SS133	Dennis	WrL	1984	Reserve
A803AWP	Dennis SS133	Dennis	WrL	1984	Reserve
A805AWP	Dennis SS133	Dennis	WrL	1984	Reserve
A806AWP	Shelvoke & Drewry WY	HCB-Angus	WrC	1984	28 Evesham
B360FAB	Dennis SS133	Dennis	WrL	1985	29 Pebworth
B361FAB	Dennis SS133	Dennis	WrL	1985	30 Broadway
B362FAB	Dennis SS133	Dennis	WrL	1985	42 Ledbury
B363FAB	Dennis SS133	Dennis	WrL	1985	49 Kington
B364FAB	Dennis SS133	Dennis	WrL	1985	44 Ross On Wye
C740ONP	Dennis SS133	Dennis	WrL	1985	Reserve
C741ONP	Dennis SS133	Dennis	WrL	1985	23 Bewdley
C742ONP	Dennis SS133	Dennis	WrL	1985	22 Stourport on Severn
C743ONP	Dennis SS133	Dennis	WrL	1985	Reserve
C744ONP	Dennis SS133	Dennis	WrL	1985	54 Bromyard

Hereford & Worcester's latest aerial appliance is L301UWP, a Volvo FL7 which has bodywork by Angloco- and the only Volvo with this brigade. The appliance has inherited a Simon Snorkel SS263 boom which was originally new to a Ford appliance in 1977. *Angloco Limited*

D908BUY	Dennis SS135	John Dennis	WrL	1987	52 Leominster
D909BUY	Dennis SS135	John Dennis	WrL	1987	51 Kingsland
D910BUY	Dennis SS135	John Dennis	WrL	1987	46 Hereford
D911BUY	Dennis SS135	John Dennis	WrT	1987	53 Tenbury Wells
D912BUY	Dennis SS135	John Dennis	WrL	1987	48 Eardisley
E124FCJ	GMC Chevrolet K30 4x4	HCB-Angus	ET	1987	25 Bromsgrove
Pod	Penman Multilift	B.A. Training Unit	BA(T)	1988	33 Droitwich
E116FCJ	Dennis SS135	Carmichael	WrL	1988	47 Eywas Harold
E117FCJ	Dennis SS135	Carmichael	WrL	1988	27 Redditch
E118FCJ	Dennis SS135	Carmichael	WrL	1988	43 Fownhope
E119FCJ	Dennis SS135	Carmichael	WrL	1988	45 Whitchurch
E120FCJ	Dennis SS135	Carmichael	WrL	1988	25 Bromsgrove
F762RAB	Leyland Freighter T45	Leyland-DAF	PM	1988	Training Centre
G653HLS	Steyr-Puch Pinzgauer	Outreach	L6V	1989	46 Hereford
G177GAB	Austin Maestro Van	Austin	FIU	1990	26 Droitwich
G777FWP	Steyr-Puch Pinzgauer	Mountain Range	L6V	1990	41 Malvern
G779FWP	Dennis DF237	Carmichael/Magirus DLK30	TL	1990	46 Hereford
G781FWP	GMC 3500 4x4	HCB-Angus	ET	1990	21 Worcester
G782FWP	Dennis DFS137	Carmichael/Magirus DLK30	TL	1990	21 Worcester
G783FWP	Dennis SS133	John Dennis	WrL	1990	44 Ross On Wye
G784FWP	Dennis SS133	John Dennis	WrL	1990	21 Worcester
G785FWP	Dennis SS133	John Dennis	WrL	1990	50 Leintwardine
G786FWP	Dennis SS133	John Dennis	WrL	1990	26 Droitwich
G787FWP	Dennis SS133	John Dennis	WrL	1990	31 Pershore
G788FWP	Dennis SS135	John Dennis	WrL	1990	28 Evesham
G789FWP	Crayford Engineering Argocat 8x8	Crayford Engineering	L8V	1990	41 Malvern
H592TUY	Ford Transit	Mountain Range	L/WrT	1991	45 Whitchurch
H912TUY	Dennis SS235	John Dennis	ISU	1991	26 Droitwich
H913TUY	GMC Chevrolet 3500 4x4	HCB-Angus (1987)	ET	1991	46 Hereford
H914TUY	Crayford Engineering Argocat 8x8	Crayford Engineering	L8V	1991	22 Stourport On Severn
J667BUY	Dennis SS237	John Dennis	WrL	1991	54 Bromyard
J668BUY	Dennis SS237	John Dennis	WrL	1991	53 Tenbury Wells
J669BUY	Dennis SS237	John Dennis	WrL	1991	24 Kidderminster
J670BUY	Dennis SS237	John Dennis	WrL	1991	32 Upton upon Severn
J671BUY	Leyland-DAF FA60	Dairy Products	WrC	1991	24 Kidderminster
K801KAB	Dennis Rapier TF202	John Dennis	WrL	1992	41 Malvern
K802KAB	Dennis SS237	John Dennis	WrL	1991	42 Ledbury
K803KAB	Dennis SS237	John Dennis	WrL	1991	52 Leominster
L292UWP	Dennis Rapier TF202	John Dennis	WrL	1993	26 Droitwich
L293UWP	Dennis Rapier TF202	John Dennis	WrL	1993	25 Bromsgrove
L294UWP	Dennis Rapier TF202	John Dennis	WrL	1993	21 Worcester
L295UWP	Dennis Rapier TF202	John Dennis	WrL	1993	21 Worcester
L301UWP	Volvo FL7	Angloco / Simon Snorkel SS263	HP	1993	27 Redditch
L961YAB	Dennis Rapier TF202	John Dennis	WrL	1994	24 Kidderminster
L962YAB	Dennis Rapier TF202	John Dennis	WrL	1994	24 Kidderminster
L963YAB	Dennis Rapier TF202	John Dennis	WrL	1994	27 Redditch
L964YAB	Dennis Rapier TF202	John Dennis	WrL	1994	27 Redditch
M281KWP	Dennis Sabre TSD233	John Dennis	WrL	1995	28 Evesham
M282KWP	Dennis Sabre TSD233	John Dennis	WrL	1995	41 Malvern
M283KWP	Dennis Sabre TSD233	John Dennis	WrL	1995	46 Hereford
M284KWP	Dennis Sabre TSD233	John Dennis	WrL	1995	46 Hereford
Boat	Tornado 5.1m	Boat & Trailer	IRB	1995	21 Worcester

Note: The bodywork on H913TUY was built in 1987; G653HLS was formerly a demonstrator, acquired in 1992

HERTFORDSHIRE FIRE & RESCUE SERVICE

Hertfordshire Fire & Rescue Service, Old London Road, Hertford SG13 7LD

W001	DJH523	Leyland FT3	Leyland	PE	1937	Preserved at C24 Hertford
W002		Shand Mason	Horse drawn steamer appliance	P	1901	Preserved at C23 Stevenage
W054	KNK442V	Ford A0610	Longwell Green	GPV	1979	FSNBF
W057	D315PVS	Ford Transit	Ford	PCV	1987	Training Centre
W078	C342WHA	Dennis F127	Saxon Sanbec/Simon SS263	HP	1985	C23 Stevenage
W080	E568DMJ	Dennis F127	Saxon Sanbec/Simon SS263	HP	1988	A09 Watford
W082	G726WBH	Mercedes-Benz 917AF	HCB-Angus	ERT	1989	B18 Hatfield
W083	G314WNM	Dennis F127	Saxon Sanbec/Simon SS263	HP	1990	B11 St.Albans
W086	J960CRB	Mercedes-Benz 814D	Leicester Carriage	CU	1991	B18 Hatfield
W087	J783PKX	Iveco-Ford Cargo O915	Leicester Carriage	GPV	1992	C23 Stevenage
W088	J899MRO	Land Rover Defender 110	Land Rover	L4V	1991	B19 Welwyn Garden City
W089	J309OTM	Land Rover Defender 110	Land Rover	L4V	1991	A01 Hemel Hempstead
W090	J782PKX	Iveco-Ford Cargo O915	Leicester Carriage	SU	1992	B11 St.Albans
W091	J581SNK	Land Rover Defender 110	Land Rover	L4V	1992	C23 Stevenage
W093	K873BBH	Iveco-Ford Cargo 100E18	Leicester Carriage	HMU	1993	A01 Hemel Hempstead
W094	K57CRO	Ford Transit	Ford	PCV	1993	B18 Hatfield
W095	K58CRO	Ford Transit	Ford	PCV	1993	Training Centre
W097	N335PNH	Iveco-Ford Cargo 100E18	Leicester Carriage	DCU	1995	C32 Hitchen
W098	N502VNH	Iveco-Ford Cargo 100E18	Leicester Carriage	DCU	1996	A07 Rickmansworth
W615	A317NRO	Dennis SS133	Dennis	WrL	1983	B Reserve
W616	A318NRO	Dennis SS133	Dennis	WrL	1983	C Reserve
W617	A319NRO	Dennis SS133	Dennis	WrL	1983	B Reserve
W619	A253HPE	Dennis SS135	Dennis	WrE	1984	A09 Watford (Cadet scheme / Preserved)
W620	A252HPE	Dennis SS135	Dennis	WrL	1984	Training Centre
W621	A251HPE	Dennis SS135	Dennis	WrL	1984	A Reserve
W622	A250HPE	Dennis SS135	Dennis	WrL	1984	C Reserve
W623	B158MPM	Dennis SS135	Dennis	WrL	1985	C Reserve
W624	B157MPM	Dennis SS135	Dennis	WrL	1985	A Reserve
W625	B125CVH	Shelvoke & Drewry WX	Angloco	WrL	1985	Training Centre
W626	B126CVH	Shelvoke & Drewry WX	Angloco	WrL	1985	Training Centre
W627	B127CVH	Shelvoke & Drewry WX	Angloco	WrL	1985	Training Centre
W628	B128CVH	Shelvoke & Drewry WX	Angloco/Hertfordshire FS	CU/ERT	1985	B Reserve
W629	C98RPJ	Dennis DS151	Saxon Sanbec	WrT	1985	A08 Bushey
W630	C99RPJ	Dennis DS151	Saxon Sanbec	WrT	1985	Training Centre
W631	C101RPJ	Dennis DS151	Saxon Sanbec	WrT	1985	B20 Welwyn
W632	C102RPJ	Dennis DS151	Saxon Sanbec	WrT	1985	B21 Wheathampstead
W633	C103RPJ	Dennis DS151	Saxon Sanbec	WrT	1985	C29 Buntingford
W634	D932TBM	Dennis DS151	Saxon Sanbec	WrT	1987	C26 Much Hadham
W635	D931TBM	Dennis DS151	Saxon Sanbec	WrT	1987	A05 Bovingdon
W636	D930TBM	Dennis DS151	Saxon Sanbec	WrT	1987	B13 Radlett
W637	D929TBM	Dennis DS151	Saxon Sanbec	WrT	1987	B17 Hoddesdon
W638	D928TBM	Dennis DS151	Saxon Sanbec	WrT	1987	A03 Tring
W639	E804BMJ	Dennis DS153	Saxon Sanbec	WrT	1987	B12 Redbourn
W640	E805BMJ	Dennis DS153	Saxon Sanbec	WrT	1987	B22 Harpenden
W641	E806BMJ	Dennis DS153	Saxon Sanbec	WrT	1987	C24 Hertford
W642	E807BMJ	Dennis DS153	Saxon Sanbec	WrT	1987	A02 Markyate
W643	E808BMJ	Dennis DS153	Saxon Sanbec	WrT	1987	C27 Sawbridgeworth
W644	F656LNM	Dennis SS235	Carmichael	WrL	1988	B16 Cheshunt
W645	F657LNM	Dennis SS235	Carmichael	WrL	1988	A04 Berkhamsted
W646	F658LNM	Dennis SS235	Carmichael	WrL	1988	A01 Hemel Hempstead
W647	F659LNM	Dennis SS235	Carmichael	WrL	1988	B Reserve
W648	F660LNM	Dennis SS235	Carmichael	WrL	1988	C28 Bishops Stortford
W649	F661LNM	Dennis SS235	Carmichael	WrL	1988	C30 Royston
W650	G315WNM	Dennis SS235	Carmichael	WrL	1990	C25 Ware
W651	G316WNM	Dennis SS235	Carmichael	WrL	1990	B19 Welwyn Garden City
W652	G317WNM	Dennis SS235	Carmichael	WrL	1990	A06 Kings Langley
W653	G318WNM	Dennis SS235	Carmichael	WrL	1990	B16 Cheshunt
W654	K291XMJ	Dennis SS235	Fulton & Wylie	WrL	1992	C23 Stevenage
W655	K292XMJ	Dennis SS235	Fulton & Wylie	WrL	1992	C32 Hitchin
W656	K293XMJ	Dennis SS235	Fulton & Wylie	WrL	1992	B19 Welwyn Garden City
W657	K294XMJ	Dennis SS235	Fulton & Wylie	WrL	1992	C31 Baldock & Letchworth

Hertfordshire's fifteen Dennis D5 Water Tenders are mostly allocated to retained fire stations and are represented by W631 (C101RPJ) seen at Welwyn. On the other side of the town is Welwyn Garden City with a wholetime station that sports two Dennis SS235 WrLs and a Land Rover. *Colin Lloyd*

W658	K295XMJ	Dennis SS235	Fulton & Wylie	WrL	1992	A09 Watford
W659	K26APP	Dennis SS235	John Dennis	WrL	1992	A10 Garston
W660	K25APP	Dennis SS235	John Dennis	WrL	1992	B14 Borehamwood
W661	K24APP	Dennis SS235	John Dennis	WrL	1992	C24 Hertford
W662	K21APP	Dennis SS235	John Dennis	WrL	1992	C28 Bishops Stortford
W663	K23APP	Dennis SS235	John Dennis	WrL	1992	C30 Royston
W664	L177JRO	Mercedes-Benz 1324	John Dennis	WrL	1994	B18 Hatfield
W665	L669NBH	Dennis SS241	John Dennis	WrL	1994	C23 Stevenage
W666	L670NBH	Dennis SS241	John Dennis	WrL	1994	A09 Watford
W667	L667NBH	Dennis SS241	John Dennis	WrL	1994	A01 Hemel Hempstead
W668	L668NBH	Dennis SS241	John Dennis	WrL	1994	B11 St Albans
W669	N749DUR	Dennis Sabre TSD233	John Dennis	WrL	1995	B15 Potters Bar
W670	N750DUR	Dennis Sabre TSD233	John Dennis	WrL	1995	A07 Rickmandsworth
W671	N751DUR	Dennis Sabre TSD233	John Dennis	WrL	1995	B11 St Albans
W672	N851FBM	Dennis Sabre TSD233	John Dennis	WrL	1995	C32 Hitchen
W673	N852FBM	Dennis Sabre TSD233	John Dennis	WrL	1995	C31 Boldock & Letchworth
W674	N927KMJ	Dennis Sabre TSD233	John Dennis	WrL	1996	C30 Royston
W675	N928KMJ	Dennis Sabre TSD233	John Dennis	WrL	1996	C24 Hertford
W676	N929KMJ	Dennis Sabre TSD233	John Dennis	WrL	1996	A01 Hemel Hempstead
W677	N930KMJ	Dennis Sabre TSD233	John Dennis	WrL	1996	C28 Bishops Stortford
W678	N931KMJ	Dennis Sabre TSD233	John Dennis	WrL	1996	B16 Cheshunt

Notes: W054 was previously a CIU; W628 was previously a WrL, and can be used as either an ET or CU. W664 was previously a demonstrator for Mercedes-Benz and is now named 'Jet'.

HIGHLAND & ISLANDS FIRE BRIGADE

Highland & Islands Fire Brigade, 16 Harbour Road, Longman West,
Inverness IV1 1TB

KSA363	Dennis F8	Dennis	WrT	1955	Preserved
POW647G	Muskeg Bombardier	Tracked All-terrain vehicle	ATV	1969	Inverness
XLS280T	Dodge G1313	Fulton & Wylie	WrL	1979	Recovery
YAS694V	Bedford CF	Fulton & Wylie	LFA	1980	Workshops
AST511W	Dodge G13C	Fulton & Wylie	WrL	1981	Workshops
AST982W	Dodge G13C	Fulton & Wylie	WrL	1981	Reserve
AST983W	Dodge G13C	Fulton & Wylie	WrL	1981	Reserve
EST458X	Dodge G13C	Fulton & Wylie	WrL	1982	Reserve
GAS233X	Dodge G13C	Fulton & Wylie	WrL	1982	Reserve
GAS234X	Dodge G13C	Fulton & Wylie	WrL	1982	Reserve
JST469Y	Dodge G13C	Fulton & Wylie	WrL	1982	B6 Stornoway
LAS160Y	Ford Cargo 0811	Wilcox-Seadyke	CU	1983	A1 Inverness
MST681Y	Ford Escort 35 van	Highland & Islands FB	ULFA	1983	Balintore
MST683Y	Ford Escort 35 van	Highland & Islands FB	ULFA	1983	Reserve
MST687Y	Ford Escort 35 van	Highland & Islands FB	ULFA	1983	Durness
A861OAS	Bedford CF2	Fulton & Wylie	LFA	1983	A51 Strontian
A977PAS	Ford Escort 55 van	Highland & Islands FB	ULFA	1984	B42 Achiltibuie
A982PST	Ford Escort 35 van	Highland & Islands FB	ULFA	1984	Hillswick, Shetland
A991PST	Ford Escort 35 van	Highland & Islands FB	ULFA	1984	Scalpay, Harris
A704RAS	Dodge G13C	Fulton & Wylie	WrL	1984	C7 Stromness, Orkney
A706RAS	Dodge G13C	Fulton & Wylie	WrL	1984	Reserve
A707RAS	Dodge G13C	Fulton & Wylie	WrL	1984	C29 Brae, Shetland
B686JSS	Dodge G16	Enterprise/H&IFB	WrC	1984	A1 Inverness
B816UST	Dodge G16	Fulton & Wylie	WrL	1985	Kirkwall, Orkney

Around half of the Highlands & Islands fleet are small van based appliances for retained and volunteer stations, the other half consists of standard full sized water tenders and a few special appliances. One of the 'specials' is B686JSS, the Inverness based Water Carrier. *Andrew Fenton*

B817UST	Bedford CF2	Fulton & Wylie	LFA	1985	Lochcarron
B679VST	Dodge G13C	Fulton & Wylie	WrL	1985	Invergordon
B683VST	Bedford CF2	Fulton & Wylie	LFA	1985	Dunvegan, Skye
C833RNP	Dodge G13C	Carmichael Fire Chief	WrL	1985	Training Centre, Invergordan
C765XST	Dodge G13C	Fulton & Wylie	WrL	1985	Thurso
C766XST	Dodge G13C	Fulton & Wylie	WrL	1985	A3 Grantown on Spey
C230YAS	Bedford CF2	Fulton & Wylie	LFA	1985	Lerwick, Shetland
C256YAS	Dodge G13C	Fulton & Wylie	WrL	1985	B1 Dingwall
C207AAS	Dodge G13C	Fulton & Wylie	WrL	1986	C29 Brae, Shetland
C208AAS	Dodge G13C	Fulton & Wylie	WrL	1986	A7 Kinlochleven
C209AAS	Dodge G13C	Fulton & Wylie	WrL	1986	B2 Fortrose
C834AAS	Bedford CF2	Fulton & Wylie	LFA	1986	Baltasound, Shetland
C835AAS	Bedford CF2	Fulton & Wylie	LFA	1986	C41 Whalsay, Shetland
C836AAS	Bedford CF2	Fulton & Wylie	LFA	1986	Tongue
D96DST	Freight-Rover Sherpa	Fulton & Wylie	ET	1986	A1 Inverness
D669FST	Bedford CF2	Fulton & Wylie	LFA	1987	Achfary
E316GST	Ford Escort 55 van	Highland & Islands FB	ULFA	1987	C36 Mossbank, Shetland
E392HST	Renault-Dodge G13	Fulton & Wylie	WrL	1988	A3 Granton on Spey
E393HST	Renault-Dodge G13	Fulton & Wylie	WrL	1988	C7 Stromness, Orkney
E394HST	Renault-Dodge G13	Fulton & Wylie	WrL	1988	C8 Lerwick, Shetland
E159LAS	Ford Transit	Fulton & Wylie	LFA	1988	Helmsdale
E160LAS	Ford Transit	Fulton & Wylie	LFA	1988	B28 Lochmaddy, North Uist
E161LAS	Ford Transit	Fulton & Wylie	LFA	1988	B27 Lochboisdale, South Uist
F428MST	Ford Escort 55 van	Highland & Islands FB	ULFA	1988	A46 Beauly
F429MST	Ford Escort 55 van	Highland & Islands FB	ULFA	1988	Valtos, Lewis
F860RAS	Ford Transit	Fulton & Wylie	LFA	1989	B5 Ullapool
F861RAS	Ford Transit	Fulton & Wylie	LFA	1989	C38 Scalloway, Shetland
F827RST	Mercedes-Benz 1222F	Fulton & Wylie	WrL	1989	A2 Nairn
G228TAS	Mercedes-Benz 1222F	Fulton & Wylie	WrL	1989	A2 Nairn
G391VAS	Ford Escort 55 van	Highland & Islands FB	ULFA	1990	Bonar Bridge
G392VAS	Ford Escort 55 van	Highland & Islands FB	ULFA	1990	C55 Kinlochewe
G792VAS	Ford Transit	Fulton & Wylie	LFA	1990	B8 Kinlochbervie
H536AJS	Mercedes-Benz 1120F	Fulton & Wylie	WrL	1990	Stornoway, Lewis
H487AAS	Ford Transit	Fulton & Wylie	LFA	1991	B7 Benbecula, Uist
H855AST	Mercedes-Benz 1120F	Fulton & Wylie	WrL	1991	A6 Fort William
H616CAS	Mercedes-Benz 1222F	Fulton & Wylie	WrL	1991	Dornoch
H899CAS	Ford Escort 60 van	Highland & Islands FB	ULFA	1991	Flotta, Orkney
J366FAS	Mercedes-Benz 1120F	Fulton & Wylie	WrL	1991	B6 Stornoway, Lewis
J367FAS	Mercedes-Benz 1120F	Fulton & Wylie	WrL	1991	C2 Wick
J368FAS	Mercedes-Benz 1120F	Fulton & Wylie	WrL	1991	C6 Kirkwall, Orkney
J369FAS	Mercedes-Benz 1124F	Fulton & Wylie	WrL	1991	B3 Invergordon
J999EST	Mercedes-Benz 2228	F & W/Bronto Skylift 28-2Ti	HP	1992	A1 Inverness
	Crayford Engineering	Argocat 8x8 + trailer	ATV	1992	A1 Inverness
K384LAS	Land-Rover Defender 110	Highland & Islands FB	L4P	1992	A1 Inverness
K185LST	Leyland-DAF 400 V8	Highland & Islands FB	LFA	1992	Westray
K186LST	Leyland-DAF 400 V8	Highland & Islands FB	LFA	1992	Stronsay
K268MST	Mercedes-Benz 1124F	Carmichael International	WrL	1993	A10 Portree, Skye
K269MST	Mercedes-Benz 1124F	Carmichael International	WrL	1993	C1 Thurso
K270MST	Mercedes-Benz 1124F	Carmichael International	WrL	1993	A8 Mallaig
K271MST	Mercedes-Benz 1124F	Carmichael International	WrL	1993	B1 Dingwall
K997NAS	Mercedes-Benz 310D	Highland & Islands FB	LFA	1993	C37 Sandwick, South Ronaldsay
L937KJS	Mercedes-Benz 310D	Highland & Islands FB	LFA	1993	Shawbost, Lewis
L969RAS	Mercedes-Benz 310D	Highland & Islands FB	LFA	1993	Sumburgh, Shetland
L970RAS	Mercedes-Benz 310D	Highland & Islands FB	LFA	1993	Leverburgh, Harris
L216RST	Mercedes-Benz 310D	Highland & Islands FB	LFA	1993	Bettyhill
L317SAS	Mercedes-Benz 1124F	Carmichael International	WrL	1994	C4 Golspie
L318SAS	Mercedes-Benz 1124F	Carmichael International	WrL	1994	A6 Fort William
L319SAS	Mercedes-Benz 1124F	Carmichael International	WrL	1994	B4 Tain
L320SAS	Mercedes-Benz 1124F	Carmichael International	WrL	1994	Gairloch
L321SAS	Mercedes-Benz 310D	Highland & Islands FB	LFA	1994	A45 Fort Augustus
L322SAS	Mercedes-Benz 310D	Highland & Islands FB	LFA	1994	C21 Longhope, Hoy
L323SAS	Mercedes-Benz 310D	Highland & Islands FB	LFA	1994	B23 Castlebay, Barra
L324SAS	Mercedes-Benz 310D	Highland & Islands FB	LFA	1994	Lochaline
M101VAS	Mercedes-Benz 310D	Highland & Islands FB	LFA	1994	Mid Yell, Shetland
M102VAS	Mercedes-Benz 310D	Highland & Islands FB	LFA	1994	St.Margarets Hope, South Ronaldsay
M103VAS	Mercedes-Benz 310D	Highland & Islands FB	LFA	1994	Sanday
M104VAS	Mercedes-Benz 310D	Highland & Islands FB	LFA	1994	Acharacle
M105VAS	Mercedes-Benz 310D	Highland & Islands FB	LFA	1994	Raasay
M426XAS	Mercedes-Benz 1124F	Carmichael International	WrL	1995	C8 Lerwick, Shetland
M427XAS	Mercedes-Benz 1124F	Carmichael International	WrL	1995	A5 Kingussie
M428XAS	Mercedes-Benz 1124F	Carmichael International	WrL	1995	Broadford, Skye

The brigade workshops of Highland & Islands Fire Brigade convert panel vans into Lightweight Fire Appliances for Volunteer firefighters throughout the brigades considerable 12,000 square mile territory. An example of these conversions is N918YST a LDV400 V8 for Cannich. *Alistair MacDonald*

M429XAS	Volvo FL6.14	Carmichael International	WrL	1995	Benbecula
M762XAS	Mercedes-Benz 310D	Highland & Islands FB	LFA	1995	Tarbert, Harris
M763XAS	Mercedes-Benz 310D	Highland & Islands FB	LFA	1995	Lochinver
N651RJS	Iveco TurboDaily 40-10 4x4	Highland & Islands FB	LFA	1995	Skerris, Shetland
N928RJS	Iveco TurboDaily 40-10 4x4	Highland & Islands FB	LFA	1995	Foula, Shetland
N918YST	LDV 400	Highland & Islands FB	LFA	1995	Cannich
N919YST	LDV 400	Highland & Islands FB	LFA	1995	Walls, Shetland
N472BST	LDV 400	Highland & Islands FB	LFA	1996	Kilchoan
N473BST	LDV 400	Highland & Islands FB	LFA	1996	B44 Aultrea
N682BST	Mercedes-Benz 1124F	Emergency One	WrL	1996	B5 Ullapool
N683BST	Mercedes-Benz 1124F	Emergency One	WrL	1996	A4 Aviemore
N684BST	Mercedes-Benz 1124F	Emergency One	WrL	1996	C5 Lairg
N685BST	Mercedes-Benz 1124F	Emergency One	WrL	1996	A9 Kyle of Lochalsh
P519EAS	LDV Convoy	Highland & Islands FB	FT	1996	A1 Inverness
P520EAS	LDV Convoy	Highland & Islands FB	LFA	1996	Ness
P521EAS	LDV Convoy	Highland & Islands FB	LFA	1996	Eday
P140FAS	Volvo FL6.14	Emergency One	WrL	1996	A7 Kinlochleven
P141FAS	Volvo FL6.14	Emergency One	WrL	1996	A1 Inverness
P142FAS	Volvo FL6.14	Emergency One	WrL	1996	A1 Inverness
P143FAS	Volvo FL6.14	Emergency One	WrL	1996	A1 Inverness

Note: XLS280T was Ex Central Region Fire Brigade in 1992; C833RNP was originally a Carmichael demonstrator; B686JSS was originally a tipper, then fitted with a milk tank. K268MST is named Eilean na Chec.

HUMBERSIDE FIRE BRIGADE

Humberside Fire Brigade, Summergrove Way, Hessle High Road, Hull HU4 7BB

KRH582V	Dennis Delta 2 DF133	Dennis/Humberside FB	FoT	1979	B2 Bransholme
OAG10V	Dennis Delta 2 DF133	Dennis/Simon Snorkel SS220	HP	1980	D2 Goole
ERH675Y	Dennis SS133	Dennis	WrL	1982	Training Centre
ERH676Y	Dennis SS133	Dennis	WrL	1982	A Reserve
ERH677Y	Dennis SS133	Dennis	WrL	1982	C Reserve
ERH678Y	Dennis SS133	Dennis	WrL	1982	A Reserve
HKH942Y	Dennis DS151	Dennis/Humberside FB	ET/CU	1983	D1 Scunthorpe
HKH943Y	Dennis SS133	Dennis	WrL	1983	B3 Bridlington (R)
HKH944Y	Dennis SS133	Dennis	WrL	1983	C Reserve
HKH945Y	Dennis SS133	Dennis	WrL	1983	Driving School
HKH946Y	Dennis SS133	Dennis	WrL	1983	D Reserve
HKH947Y	Dennis SS133	Dennis	WrL	1983	C Reserve
A261PAG	Dennis DF133	Carmichael / Magirus DL30	TL	1984	A1 Hull Central
A162PKH	Dennis DF133	Dennis/Simon Snorkel SS220	HP	1984	C1 Grimsby Central
A163PKH	Dennis DS151	Dennis	WrT	1984	D Reserve
A164PKH	Dennis DS151	Dennis	WrT/R	1984	D4 Crowle
A165PKH	Dennis DS151	Dennis	WrT/R	1984	D7 Winterton
A166PKH	Dennis DS151	Dennis	WrT/R	1984	D5 Epworth
A167PKH	Dennis DS151	Dennis	WrT/R	1984	D6 Kirton In Lindsey
A518RRH	Ford Cargo 0813	Derwent/Humberside FB	CU/CAV	1984	A1 Hull Central
A328SAT	Ford Cargo 1013	Hartford/Humberside FB	HL	1984	C7 Barton Upon Humber
B229VRH	Dennis DF133	Dennis	FoT	1985	C4 Immingham
C550CRH	Dennis SS135	Angloco	WrL	1985	Driving School
C551CRH	Dennis SS135	Angloco	WrL	1985	B8 Patrington (R)
C552CRH	Dennis SS135	Angloco	WrL	1985	B4 Driffield (R)

Three batches of Dennis Rapier pumps have entered service with Humberside Fire Brigade. The first five appliances were to the original Rapier style, while the others have the later design as seen on M320LKH in Hull city centre. This machine is currently Hull Easts Water Tender Ladder/Rescue appliance with hydraulic rescue equipment. *Tony Wilson*

C553CRH	Dennis SS135	Angloco	WrL	1985	Driving School
C188EAG	Dennis DF135	Angloco/Metz DL30	TL	1986	D1 Scunthorpe
C870EKH	Dennis SS135	Fulton & Wylie/Humberside FB	ET	1986	A2 Hull North
D477KAG	Dennis SS135	Angloco	WrT	1986	A6 Market Weighton
D478KAG	Dennis SS135	Angloco	WrT	1986	B3 Bridlington
D479KAG	Dennis SS135	Angloco	WrT	1986	A7 Pocklington
D480KAG	Dennis SS135	Angloco	WrL	1986	D Reserve
D483KAG	Ford Cargo 0609	Nesham/Ratcliffe	GPV	1986	Ladder Workshop
E368RRH	Dennis SS135	Angloco	WrT	1987	B6 Hornsea
E369RRH	Dennis SS135	Angloco	WrT	1987	A4 Beverley
E370RRH	Dennis SS135	Angloco/Humberside FB	WrT	1987	B3 Bridlington
E371RRH	Dennis SS135	Angloco	WrT/R	1987	D9 Brigg
E439VRH	Dennis SS135	Angloco	WrT	1988	B4 Driffield
E440VRH	Dennis SS135	Angloco	WrT/R	1988	B5 Sledmore
E441VRH	Volvo FL6.14	Carmichael	WrT/R	1988	D3 Snaith
E442VRH	Volvo FL6.14	Carmichael	WrT/R	1988	C5 Waltham
F448DAG	Dennis SS135	Carmichael	WrT	1988	B7 Withernsea
F449DAG	Dennis SS135	Carmichael	WrL	1988	A4 Beverley
F450DAG	Dennis SS135	Fulton & Wylie	ET	1988	C4 Immingham
F451DAG	Dennis SS135	Carmichael	WrT	1988	A5 Brough
F452DAG	Dennis SS135	Carmichael	WrT	1988	D1 Scunthorpe
F151BAT	GMC Chevrolet Scottsorle K30	HCB-Angus	ET	1988	A Reserve
G69NAG	Dennis SS135	Carmichael	WrL/R	1989	B9 Preston
G70NAG	Dennis SS135	Carmichael	WrT/R	1989	C3 Cleethorpes
G71NAG	Dennis SS135	Carmichael	WrT/R	1989	D8 Howden
G72NAG	Dennis SS135	Carmichael	WrL/R	1989	A6 Market Weighton
G73NAG	Dennis SS135	Carmichael	WrT	1989	C1 Grimsby Central
G761LKH	Volvo FL6.17	Angloco/Metz DL30	TL	1989	B3 Bridlington
H557WRH	Dennis SS233	Carmichael	WrT	1990	C4 Immingham
H558WRH	Dennis SS233	Carmichael	WrL/R	1990	B4 Driffield
H559WRH	Dennis SS233	Carmichael	WrT/R	1990	A7 Pocklington
H561WRH	Dennis SS233	Carmichael	WrL/R	1990	B6 Hornsea
H369ARH	Dennis SS239	John Dennis	P/RT	1991	A4 Beverley
J748FRH	Dennis SS233	Carmichael	WrT/R	1991	B8 Patrington
J749FRH	Dennis SS233	Carmichael	WrL/R	1991	B7 Withernsea
J750FRH	Dennis SS233	Carmichael	WrL/R	1991	A5 Brough
J751FRH	Dennis SS233	Carmichael	WrT	1991	C2 Cromwell Road, Grimsby
K469PAG	Volvo FL10	Bedwas/Simon Snorkel ST290-S	ALP	1992	A3 Hull West
K470PAG	Dennis SS239	John Dennis	WrT	1993	A1 Hull Central
K471PAG	Dennis SS239	John Dennis	WrT	1993	B1 Hull East
K472PAG	Dennis SS239	John Dennis	WrT	1993	C1 Grimsby Central
K473PAG	Dennis SS239	John Dennis	WrT	1993	A2 Hull North
L236CAG	Dennis Rapier TF203	John Dennis	WrL/RT	1994	B3 Bridlington
L237CAG	Dennis Rapier TF203	John Dennis	WrT	1994	A3 Hull West
L238CAG	Dennis Rapier TF203	John Dennis	WrT	1994	D2 Goole
L239CAG	Dennis Rapier TF203	John Dennis	WrT	1994	D1 Scunthorpe
L240CAG	Dennis Rapier TF203	John Dennis	WrT/R	1994	C7 Barton-upon-Humber
M318LKH	Dennis Rapier TF203	John Dennis	WrL/RT	1995	A2 Hull North
M319LKH	Dennis Rapier TF203	John Dennis	WrL/R	1995	A1 Hull Central
M320LKH	Dennis Rapier TF203	John Dennis	WrL/R	1995	B1 Hull East
M321LKH	Dennis Rapier TF203	John Dennis	WrL/R	1995	C2 Cromwell Rd, Grimsby
M322LKH	Dennis Rapier TF203	John Dennis	WrL/R	1995	C1 Grimsby Central
N922ERH	Dennis Rapier R411	Carmichael International	WrL/RT	1996	C4 Immingham
N923ERH	Dennis Rapier R411	Carmichael International	WrL/R	1996	B2 Bransholme
N924ERH	Dennis Rapier R411	Carmichael International	WrL/RT	1996	D1 Scunthorpe
N925ERH	Dennis Rapier R411	Carmichael International	WrL/RT	1996	D2 Goole
N926ERH	Dennis Rapier R411	Carmichael International	WrL/RT	1996	A3 Hull West
N	Iveco EuroCargo	G C Smith	CU	1996	

ISLE OF MAN FIRE & RESCUE SERVICE

Isle of Man Fire & Rescue Service, Peel Road, Douglas, Isle of Man

999MMN	ERF 84RF	ERF Firefighter	WrT	1977	Reserve
999UMN	Dennis R61	Dennis	WrT	1979	Reserve
OMN999	Dennis DS151	Dennis	WrT	1981	2 Laxey
999NMN	Land Rover Series III 109	Pilcher Green	L4P	1981	4 Kirk Michael
TMN999	Dennis DS151	Dennis	WrL	1983	6 Rushen (Port Erin)
999PMN	Dennis DS151	Dennis	WrT	1984	7 Castletown
BMN999A	Mercedes-Benz Unimog 1300L	Fulton & Wylie	L4P	1987	2 Laxey
MAN999B	Volvo FL10	Nova Scotia	Fo/HL	1988	1 Douglas
MAN999U	Volvo FL6.17	Angloco/Metz DL30M	TL	1988	1 Douglas
BMN999L	Steyr Pinzgauer	Saxon Sanbec	L6P	1989	5 Peel
BMN999P	Volvo FL6.14	Fulton & Wylie	WrL	1990	1 Douglas
BMN999R	Steyr Pinzgauer	Saxon Sanbec	L6P	1990	3 Ramsey
BMN999T	Steyr Pinzgauer	Saxon Sanbec	L/ET	1991	7 Castletown
BMN999U	Mercedes-Benz 917AF	Carmichael	P/FoT	1990	1 Douglas
BMN999W	Nissan 4x4	Truckman	L4V	1991	Headquarters
CMN999E	Steyr Pinzgauer	Saxon Sanbec	L6P	1991	6 Rushen (Port Erin)
FMN999	Land Rover Discovery	Land Rover	L4V	1991	1 Douglas
NMN999	Volvo FL6.17	Carmichael/Simon Snorkel SS220	HP	1992	1 Douglas
CMN999D	Mercedes-Benz 1120AF	Carmichael	WrL	1992	3 Ramsey
999GMN	Volvo FL6.14	Penman/Outreach/Palfinger PK8000ET		1992	1 Douglas
999CMN	Volvo FL6.14	Carmichael International	WrL	1993	5 Peel
999WMN	Volvo FL6.14	Carmichael International	WrL	1993	1 Douglas
KMN999	Volvo FL6.14	Carmichael International	WrL	1994	3 Ramsey
PMN999	Mercedes-Benz 1124AF	Carmichael International	WrL	1994	4 Kirk Michael
WMN999	Mercedes-Benz 1124AF	Carmichael International	WrL	1994	7 Castletown
HMN999	Steyr-Daimler-Puch Pinzgauer	Saxon Sanbec	L/ET	1994	3 Ramsey
	Steyr-Daimler-Puch Pinzgauer	Saxon Sanbec	L6P	1996	

Heavy investment in the Isle of Man Fire & Rescue service has seen a modernisation of the fleet during the last decade. Delivered in 1993, 999CMN is a Volvo FL6 water-tender-ladder with Carmichael International bodywork is now based at Peel fire station. *Bill Murray*

ISLE OF WIGHT FIRE & RESCUE SERVICE

Isle of Wight Fire & Rescue Service, South Street, Newport PO30 1JQ

SDL999S	Bedford TKG	HCB-Angus CSV	WrL	1977	Reserve
RDL502S	Bedford TKG	HCB-Angus CSV	WrL	1978	Reserve
YDL999T	Bedford TKG	HCB-Angus CSV	WrL	1979	Reserve
DDL999V	Bedford TKG	IoWFB / BRS	FSU	1980	3 East Cowes
EDL999V	Bedford TKG	HCB-Angus CSV	WrT	1980	7 Shanklin
JDL997W	Bedford KM	IoWFB	WrC	1980	4 Ryde
JDL998W	Bedford TKG	HCB-Angus CSV	WrL	1980	6 Sandown
JDL999W	Bedford TKG	HCB-Angus CSV	WrT	1980	6 Sandown
PDL528X	Bedford TKG	HCB-Angus CSV	WrL	1981	3 East Cowes
BPF588Y	Dennis DS151	Dennis	WrL	1982	5 Bembridge
VDL190Y	Bedford TL400D	HCB-Angus / IoWFB	P	1983	Driving School
A286CDL	Bedford TK1630	Buckingham Tankers	WrC	1983	9 Freshwater
A854EDL	Dennis DS151	Dennis	WrL	1984	10 Yarmouth
A855EDL	Dennis DS151	Dennis	WrL	1984	4 Ryde
C165UDL	Dennis DS151	HCB-Angus	WrL	1986	2 Cowes
D324CDL	Dennis DS151	HCB-Angus	WrT	1987	2 Cowes
E826GDL	Dennis DS153	HCB-Angus	WrL	1987	8 Ventnor
E432HDL	Freight-Rover Sherpa	Freight-Rover	PCV	1988	1 Newport
E753JDL	Scania 92M	Angloco / Bronto Skylift 22-2Ti	ALP	1988	1 Newport
F258SDL	Dennis DS153	Saxon Sanbec	WrL	1989	4 Ryde
G304WDL	Land Rover 110	Land Rover/IoWFB	L4V	1989	1 Newport
G917OTO	Ford Transit	Ford	BAT	1990	4 Ryde
H311CDL	Leyland-DAF 400	Leyland-DAF / IoWFB	BAT	1991	Workshops
H681EDL	Mercedes-Benz 917AF	Sparshatts	RT	1991	1 Newport
J95LDL	Mercedes-Benz 1120AF	Sparshatts	WrL	1992	7 Shanklin
J96LDL	Mercedes-Benz 1120AF	Sparshatts	WrL	1992	8 Ventnor
L691XDL	Mercedes-Benz 1726	Angloco / Metz DLK30PLC	TL	1994	1 Newport
L531YDL	Mercedes-Benz 1120AF	Carmichael International	WrL	1995	1 Newport
M963HDL	Mercedes-Benz 1124AF	Carmichael International	WrL	1995	1 Newport
M964HDL	Mercedes-Benz 1124AF	Carmichael International	WrL	1995	1 Newport
M344HDL	Mercedes-Benz 609D	Mercedes-Benz/IoWFS	FSU	1995	1 Newport
N477PDL	Mercedes-Benz 814D	Saxon Sanbec/IoWFS	CU	1995	1 Newport
Trailer	?	Corrosives Tanker	CT	19..	1 Newport

Notes: VDL190Y was originally a demonstrator for HCB-Angus; BPF588Y was a Dennis demonstrator.

Recent years have seen the Isle of Wight Fire & Rescue service procure many Mercedes-Benz appliances. Two 1124AF all wheel drive Water Tender Ladders built by Carmichael International, were new in 1995, both were issued to the only whole time manned fire station on the Island - Section 1 Newport. This view depicts 5045 (M964HDL) at home. *Keith Grimes*

KENT FIRE BRIGADE

Kent Fire Brigade, The Godlands, Tovil, Maidstone ME15 6XB

109	M794PKJ	Vauxhall Brava 4x4	Truckman/Whitacres	CU	1994	S60 Maidstone
110	M797PKJ	Vauxhall Brava 4x4	Truckman/Whitacres	CU	1994	E80 Canterbury
114	M795PKJ	Vauxhall Brava 4x4	Truckman/Whitacres	CU	1994	Workshops
115	M796PKJ	Vauxhall Brava 4x4	Truckman/Whitacres	CU	1994	N43 Medway
116	F96DKN	Land Rover 127 V8	Land Rover	LR	1988	N48 Sheppy
117	F97DKN	Land Rover 127 V8	Land Rover	LR	1988	N39 Strood
118	F98DKN	Land Rover 127 V8	Land Rover	LR	1988	E91 Deal
119	K819SKE	Dennis SS239	HCB-Angus	WrT	1992	E91 Deal
121	A504FPD	Dennis DS151	Dennis	WrT	1984	N42 Gillingham
124	GKO224	Leyland Comet FT4a	Braidwood/Merryweather	PE	1939	N30 Dartford (Preserved)
125	C931BKJ	Dennis SS133	Dennis	RWrL	1985	Reserve
126	C932BKJ	Dennis SS133	Dennis	WrT	1985	E17 St. Margarets
127	C933BKJ	Dennis SS133	Dennis	WrT	1986	Reserve
128	C934BKJ	Dennis SS133	Dennis	WrL	1986	S70 Edenbridge
129	C935BKJ	Dennis SS133	Dennis	WrT	1986	S12 Chilham
130	C936BKJ	Dennis SS133	Dennis	WrT	1986	N41 Chatham
131	C937BKJ	Dennis SS133	Carmichael	RWrL	1985	Training Centre
132	F999MAR	Scania G93M-250	John Dennis	WrT	1989	E19 Folkestone
133	F996FKO	Volvo FL6.14	HCB-Angus	WrT	1989	S13 Wye
135	H196FKP	Dennis Rapier TF202	John Dennis	WrT	1991	Training Centre
136	H197FKP	Dennis SS239	John Dennis	WrT	1991	N48 Sheppy
137	H198FKP	Dennis SS239	John Dennis	WrT	1991	N44 Rainham
138	H199FKP	Dennis SS239	John Dennis	WrL	1991	E18 Whitfield
139	H201FKP	Dennis SS239	John Dennis	WrT	1991	S67 Borough Green
140	N524XKJ	Dennis Rapier TF203	John Dennis	RWrL	1995	N48 Sheppy
141	H202FKP	Dennis SS239	John Dennis	WrT	1991	N38 Grain
142	M882OKE	Dennis Rapier TF203	John Dennis	RWrL	1995	E89 Thanet
143	M883OKE	Dennis Rapier TF203	John Dennis	RWrL	1994	E16 Dover
144	M884OKE	Dennis Rapier TF203	John Dennis	WrT	1994	S11 Ashford
145	M885OKE	Dennis Rapier TF203	John Dennis	RWrL	1994	E19 Folkestone
146	M886OKE	Dennis Rapier TF203	John Dennis	RWrL	1994	E80 Canterbury
147	M887OKE	Dennis Rapier TF203	John Dennis	RWrL	1994	E90 Ramsgate
159	TKM362X	Dennis RS133	Dennis	WrT	1982	Reserve
160	M880OKE	Dennis Rapier TF203	John Dennis	RWrL	1995	N45 Sittingbourne
161	M881OKE	Dennis Rapier TF203	John Dennis	RWrL	1994	S74 Tunbridge Wells
162	N525XKJ	Dennis Rapier TF203	John Dennis	RWrL	1995	N39 Strood
163	N526XKJ	Dennis Rapier TF203	John Dennis	RWrL	1995	E91 Deal
164	N527XKJ	Dennis Rapier TF203	John Dennis	RWrL	1995	E86 Herne Bay
165	N528XKJ	Dennis Rapier TF203	John Dennis	RWrL	1996	N84 Faversham
166	N529XKJ	Dennis Rapier TF203	John Dennis	RWrL	1995	E87 Margate
167	N530XKJ	Dennis Rapier TF203	John Dennis	RWrL	1996	S68 Sevenoaks

Three of Kent's aerial appliances are Scania 92M machines bodied by Carmichael and fitted with Bronto Skylift 28-2Ti booms. This view shows 193 (E499WKO) the Tunbridge Wells based appliance.
Gerald Mead

168	M96OKJ	Dennis Rapier TF203	John Dennis	WrT	1995	N43 Medway
169	M889OKE	Dennis Rapier TF203	John Dennis	RWrL	1994	N43 Medway
170	M890OKE	Dennis Rapier TF203	John Dennis	RWrL	1994	S11 Ashford
171	M891OKE	Dennis Rapier TF203	John Dennis	RWrL	1995	N35 Thames-side
172	M872OKE	Dennis Rapier TF203	John Dennis	WrT	1995	N30 Dartford
173	M873OKE	Dennis Rapier TF203	John Dennis	RWrL	1995	N30 Dartford
174	M874OKE	Dennis Rapier TF203	John Dennis	RWrL	1995	S65 Larkfield
175	M875OKE	Dennis Rapier TF203	John Dennis	RWrL	1995	S60 Maidstone
176	M876OKE	Dennis Rapier TF203	John Dennis	WrT	1995	S60 Maidstone
177	M877OKE	Dennis Rapier TF203	John Dennis	WrT	1995	N35 Thames-side
178	M878OKE	Dennis Rapier TF203	John Dennis	RWrL	1995	S72 Tonbridge
179	M879OKE	Dennis Rapier TF203	John Dennis	WrT	1995	E80 Canterbury
182	MKE886W	Dennis F125	Carmichael/Magirus DL30U	TL	1981	E89 Thanet
183	CWY157Y	Dennis DF133	Dennis/Magirus DL30U	TL	1982	Reserve
184	MKE888W	Dennis F125	Carmichael/Magirus DL30U	TL	1981	E19 Folkestone
186	E498WKO	Scania P92M	Carmichael/Magirus DL30U	TL	1988	E80 Canterbury
187	N551XKJ	Iveco-Magirus 120-25	GB Fire/Magirus DLK23-12cc	TL	1996	S60 Maidstone
188	P	Iveco-Magirus 120-25	GB Fire/Magirus DLK23-12cc	TL	1997	
189	P	Iveco-Magirus 120-25	GB Fire/Magirus DLK23-12cc	TL	1997	
191	D995PKN	Scania P92M	Carmichael/Bronto Skylift 28.2Ti	ALP	1987	N43 Medway
192	E497WKO	Scania 92M	Carmichael/Bronto Skylift 28.2Ti	ALP	1988	N35 Thames-Side
193	E499WKO	Scania 92M	Carmichael/Bronto Skylift 28.2Ti	ALP	1988	S74 Tunbridge Wells
194	KKK252V	Dennis F125	Dennis/Simon Snorkel SS220	HP	1980	Reserve
200	B773VKM	Dennis SS133	Dennis	WrT	1984	Reserve
201	B774VKM	Dennis SS133	Dennis	WrT	1984	S25 Hawkhurst
202	B775VKM	Dennis SS133	Dennis	WrT	1984	S62 Lenham
203	B776VKM	Dennis SS133	Dennis	WrT	1985	E87 Margate
204	B777VKM	Dennis SS133	Dennis	WrT	1985	S32 Horton Kirby
205	B778VKM	Dennis SS133	Dennis	WrT	1985	E90 Ramsgate
206	B779VKM	Dennis SS133	Dennis	WrL	1985	N49 Eastchurch
207	B780VKM	Dennis SS133	Dennis	WrT	1985	S71 Seal
208	B781VKM	Dennis SS133	Dennis	WrT	1985	Reserve
209	B782VKM	Dennis SS133	Dennis	WrT	1985	S64 Marden
210	B783VKM	Dennis SS133	Dennis	WrL	1985	N47 Queenborough
211	B784VKM	Dennis SS133	Dennis	WrT	1985	S63 Headcorn
212	B785VKM	Dennis SS133	Dennis	WrT	1985	Reserve
213	B786VKM	Dennis SS133	Dennis	WrT	1985	N33 Swancombe
214	B787VKM	Dennis SS133	Dennis	WrT	1985	S69 Westerham
215	B788VKM	Dennis SS133	Dennis	WrL	1985	S73 Paddockwood
216	B789VKM	Dennis SS133	Dennis	WrT	1985	N46 Teynham
217	B790VKM	Dennis SS133	Dennis	WrT	1985	S76 Southborough
220	K820SKE	Dennis SS239	HCB-Angus	WrL	1993	N37 Hoo
221	K821SKE	Dennis SS239	HCB-Angus	WrT	1993	E19 Folkestone
222	K822SKE	Dennis SS239	HCB-Angus	WrT	1993	E16 Dover
223	K823SKE	Dennis SS239	HCB-Angus	WrT	1993	S65 Larkfield
224	K824SKE	Dennis SS239	HCB-Angus	WrL	1993	S31 Swanley
225	K825SKE	Dennis SS239	HCB-Angus	RWrL	1993	E85 Whitstable
226	K826SKE	Dennis SS239	HCB-Angus	WrT	1993	N84 Faversham
227	K827SKE	Dennis SS239	HCB-Angus	WrT	1993	E85 Whitstable

Opposite: **Kent's N551XKJ is the first low-profile Iveco Magirus 120 for a British brigade. The lower picture shows a John Dennis-bodied Rapier of which some 28 are in the Kent fleet. Kent's 232, K832SKE, is based at Tunbridge Wells fire station and represents the Dennis SS239-type water tenders, here with a HCB-Angus body.**
Robert Hawkes/Robert Smith/Gerald Mead

228	K828SKE	Dennis SS239	HCB-Angus	WrT	1993	S72 Tonbridge
229	K829SKE	Dennis SS239	HCB-Angus	WrL	1993	E21 Hythe
230	K830SKE	Dennis SS239	HCB-Angus	WrT	1993	E86 Herne Bay
231	K831SKE	Dennis SS239	HCB-Angus	WrL	1993	E93 Sandwich
232	K832SKE	Dennis SS239	HCB-Angus	WrT	1993	S74 Tunbridge Wells
233	K833SKE	Dennis SS239	HCB-Angus	WrT	1993	E23 Lydd
234	K834SKE	Dennis SS239	HCB-Angus	WrT	1993	S68 Sevenoaks
235	K835SKE	Dennis SS239	HCB-Angus	WrL	1993	E81 Aylesham
236	K836SKE	Dennis SS239	HCB-Angus	WrL	1993	S24 Cranbrook
237	K837SKE	Dennis SS239	HCB-Angus	WrT	1993	N45 Sittingbourne
238	K838SKE	Dennis SS239	HCB-Angus	WrL	1993	S14 Charing
252	MKE912W	Dennis RS133	Dennis	WrT	1981	E22 Dymchurch
255	MKE915W	Dennis RS133	Dennis	WrT	1981	S75 Rusthall
256	MKE916W	Dennis RS133	Dennis	WrT	1981	Reserve
261	GGX426	Austin K2	Home Office	ATV	1941	N41 Chatham (Preserved)
269	MKE926W	Dennis RS133	Dennis	WrT	1981	Reserve
273	PKJ145W	Dennis RS133	Dennis	RWrL	1981	Training Centre
280	C919BKJ	Dennis SS133	Dennis	WrT	1985	E82 Sturry
281	C920BKJ	Dennis SS133	Dennis	WrT	1985	S77 Matfield
282	C921BKJ	Dennis SS133	Dennis	WrL	1985	E20 New Romney
283	C922BKJ	Dennis SS133	Dennis	WrT	1985	N36 Cliffe
284	C923BKJ	Dennis SS133	Dennis	WrT	1985	Reserve
285	C924BKJ	Dennis SS133	Dennis	WrT	1985	E19 Folkestone
286	C925BKJ	Dennis SS133	Dennis	WrT	1985	Reserve
287	C926BKJ	Dennis SS133	Dennis	WrT	1985	E92 Eastry
288	C927BKJ	Dennis SS133	Dennis	WrT	1985	E88 Westgate
289	C928BKJ	Dennis SS133	Dennis	WrT	1985	Reserve
290	C929BKJ	Dennis SS133	Dennis	RWrL	1985	Reserve
291	C930BKJ	Dennis SS133	Dennis	WrL	1985	S26 Tenterden
292	TKM493X	Dennis RS133	Dennis	WrT	1981	N40 Halling
293	TKM494X	Dennis RS133	Dennis	WrT	1981	S15 Aldington
295	TKM496X	Dennis RS133	Dennis	WrT	1981	E83 Wingham
298	TKM499X	Dennis RS133	Dennis	WrT	1981	Reserve
	P	Dennis Sabre SFD122	John Dennis	RWrL	1996	
	P	Dennis Sabre SFD122	John Dennis	RWrL	1996	
	P	Dennis Sabre SFD122	John Dennis	RWrL	1996	
	P	Dennis Sabre SFD122	John Dennis	RWrL	1996	
	P	Dennis Sabre SFD122	John Dennis	RWrL	1996	
	P	Dennis Sabre SFD122	John Dennis	RWrL	1996	
813	G313NKJ	Renault G	Boalloy	DTV	1989	Driving School
814	G314NKJ	Renault G	Boalloy	DTV	1989	Driving School
820	D182KKO	Ford Cargo D1620	Powell Duffryn Multilift	PM	1986	Reserve
821	D183KKO	Ford Cargo D1620	Powell Duffryn Multilift	PM	1986	N39 Strood
822	D184KKO	Ford Cargo D1620	Powell Duffryn Multilift	PM	1986	Reserve
823	E581UKP	Ford Cargo D1620	Powell Duffryn Multilift	PM	1987	Reserve
824	E353VKM	Ford Cargo D1620	Powell Duffryn Multilift	PM	1987	N84 Faversham
825	E352VKM	Ford Cargo D1620	Powell Duffryn Multilift	PM	1987	Reserve
826	E831XKE	Ford Cargo D1620	Powell Duffryn Multilift	PM	1988	E16 Dover
827	E832XKE	Ford Cargo D1620	Powell Duffryn Multilift	PM	1988	N45 Sittingbourne
828	E833XKE	Ford Cargo D1620	Powell Duffryn Multilift	PM	1988	S72 Tonbridge
829	G183OKE	Ford Cargo D1720	Powell Duffryn Multilift	PM	1989	Reserve
851	G888TKN	Iveco-Ford Cargo 1720	Reynolds Boughton	BWrC	1990	S11 Ashford
852	G826SKP	Iveco-Ford Cargo 1720	Reynolds Boughton	BWrC	1990	S65 Larkfield
898	UDT177S	Leyland Atlantean AN68/1R	East Lancashire	ExU	1978	? (Fire Safety)
901	Boat	Dell Quay Dory Rigid Raider	Rescue Boat & trailer	EBt	1987	N48 Sheppy
902	Boat	Dell Quay Dory Rigid Raider	Rescue Boat & trailer	EBt	1987	N39 Strood
903	Boat	Zodiac	Rescue Boat & trailer	EBt	1987	N48 Sheppy
904	Boat	Avon	Rescue Boat & trailer	EBt	1987	S11 Ashford
905	Boat	Avon Inlatable	Rescue Boat & trailer	IWBt	1987	S65 Larkfield
	Pod	Powell Duffryn	Breathing Apparatus	BAU	19?	
	Pod	Powell Duffryn	Channel Tunnel Module	ETU	19?	E16 Dover
	Pod	Powell Duffryn	Foam Carrier	FoT	19?	
	Pod	Powell Duffryn	Foam Carrier	FoT	19?	
	Pod	Powell Duffryn	Hose Layer	HL	19?	
	Pod	Powell Duffryn	Hose Layer	HL	19?	
	Pod	Powell Duffryn	Control Module	ICU	19?	
	Pod	Powell Duffryn	Control Module	ICU	19?	
	Pod	Powell Duffryn	Incident Support Module	ISU	19?	
	Pod	Powell Duffryn	Incident Support Module	ISU	19?	
	Pod	Powell Duffryn	Incident Support Module	ISU	19?	
	Pod	Powell Duffryn	General Purpose	GPV	19?	Training Centre

Lancashire County Fire Brigade, Garstang Road, Fulwood, Preston PR2 3LH

	u/r					
264	HCB500	Shand Mason	Shand Mason	P	1880	Headquarters (Preserved)
		Dennis F15	Dennis	PE	1956	B71 Blackburn (Preserved)
30001	L949EFV	Leyland-DAF FA45.160	Fosters Commercials	ISU	1994	A11 Lancaster (R)
30002	L950EFV	Leyland-DAF FA45.160	Fosters Commercials	ISU	1994	B70 Accrington
30003	L951EFV	Leyland-DAF FA45.160	Fosters Commercials	ISU	1994	A11 Lancaster
30004	L952EFV	Leyland-DAF FA45.160	Fosters Commercials	ISU	1994	C54 Chorley
30005	L953EFV	Leyland-DAF FA45.160	Fosters Commercials	ISU	1994	A37 South Shore, Blackpool
30007	C773OCW	Dodge G16	Angloco/Metz DLK30	TL	1986	A31 Bispham
30018	G774GFR	DAF 1900	Angloco/Metz DL30	TL	1989	B90 Burnley
30019	K156SCK	Leyland Freighter T45-180	Fulton & Wylie/Emergency One	PL	1992	C56 Skelmersdale
30020	K157SCK	Leyland-DAF 60-180	Fulton & Wylie/Emergency One	PL	1992	C57 Penwortham
30021	K158SCK	Leyland-DAF 60-210	Fulton & Wylie/Emergency One	PL	1992	C50 Preston
30022	K159SCK	Leyland-DAF 60-210	Fulton & Wylie/Emergency One	PL	1992	C50 Preston
30023	K160SCK	Leyland-DAF 60-210	Fulton & Wylie/Emergency One	PL	1992	B90 Burnley
30024	K161SCK	Leyland-DAF 60-210	Fulton & Wylie/Emergency One	PL	1992	B90 Burnley
30025	K162SCK	Leyland-DAF 60-210	Fulton & Wylie/Emergency One	PL	1992	B71 Blackburn
30026	K163SCK	Leyland-DAF 60-210	Fulton & Wylie/Emergency One	PL	1992	B71 Blackburn
30027	D100TRN	Leyland Freighter T45-180	Fulton & Wylie	PL	1987	C54 Chorley
30028	D101TRN	Leyland Freighter T45-180	Fulton & Wylie	PL	1987	C59 Longridge
30029	D102TRN	Leyland Freighter T45-180	Fulton & Wylie	PL	1987	B72 Great Harwood
30030	D103TRN	Leyland Freighter T45-180	Fulton & Wylie	PL	1987	C55 Leyland (R)
30031	D104TRN	Leyland Freighter T45-180	Fulton & Wylie	PL	1987	B92 Padiham
30032	D105TRN	Leyland Freighter T45-180	Fulton & Wylie	PL	1987	B77 Oswaldtwistle
30033	D106TRN	Leyland Freighter T45-180	Fulton & Wylie	PL	1987	B94 Nelson
30034	D107TRN	Leyland Freighter T45-180	Fulton & Wylie	PL	1987	B95 Earby
30035	D108TRN	Leyland Freighter T45-180	Fulton & Wylie	PL	1987	B74 Rawtenstall
30036	D109TRN	Leyland Freighter T45-180	Fulton & Wylie	PL	1987	A32 Fleetwood
30037	D110TRN	Leyland Freighter T45-180	Fulton & Wylie	PL	1987	C58 Tarleton
30038	D111TRN	Leyland Freighter T45-180	Fulton & Wylie	PL	1987	B70 Accrington
30039	E112GCW	Bedford TL860	Fulton & Wylie	ET	1987	Training Centre
30048	M701UHG	Leyland-DAF FA60.180	Multilift	PM	1994	B90 Burnley
30049	M702UHG	Leyland-DAF FA60.180	Multilift	PM	1994	C50 Preston
30050	M703UHG	Leyland-DAF FA60.180	Multilift	PM	1994	C50 Preston
30057	VFV729V	Bedford TKG	Cheshire Fire Engineering	PL	1980	Training Centre
30064	Pod	Ro2A	Foam Tender	FoT	1980	B90 Burnley
30065	Pod	Ro2B	Foam Tender	FoT	1980	B90 Burnley
30067	Pod	Ro1	Chemical Incident Unit	CIU	1980	C50 Preston
30068	Pod	Ro2C	Foam Tender	FoT	1980	C50 Preston
30069	Pod	Ro3	General Purpose Vehicle	GPV	1980	B71 Blackburn
30070	Pod	Fosters Commercials	Damage Control	DCU	1995	B90 Burnley
30071	Pod	Ro5	Breathing Apparatus Tender	BAT	1980	B90 Burnley
30072	Pod	Ro6	Major Incident Support	MISU	1995	C50 Preston
30101	H801YRN	Leyland-DAF T45-180	Bedwas	DTV	1991	Training Centre
30102	H802YRN	Leyland-DAF T45-180	Bedwas	DTV	1991	Training Centre
30103	K803SRN	Leyland-DAF T60-180	Bedwas	DTV	1992	Training Centre
30108	GFR598W	Dodge G1613	Carmichael/Magirus DL30	TL	1981	Reserve
30109	NCK935X	Dodge G1613	Carmichael/Magirus DL30	TL	1981	A12 Morcambe
30110	FBV831Y	Dennis DF131	Carmichael/Magirus DL30	TL	1983	B71 Blackburn
30111	FBV832Y	Dennis DF131	Carmichael/Magirus DL30	TL	1983	C50 Preston
30114	F128JEO	Leyland Freighter T45-180	Mountain Range	PL	1988	B72 Great Harwood
30115	F129JEO	Leyland Freighter T45-180	Mountain Range	PL	1988	A11 Lancaster
30116	F130JEO	Leyland Freighter T45-180	Mountain Range	PL	1988	C51 Ormskirk
30117	F131JEO	Leyland Freighter T45-180	Mountain Range	PL	1988	Reserve
30118	F132JEO	Leyland Freighter T45-180	Mountain Range	PL	1988	A32 Fleetwood
30119	F133JEO	Leyland Freighter T45-180	Mountain Range	PL	1988	B96 Colne
30120	F134JEO	Leyland Freighter T45-180	Mountain Range	PL	1988	B73 Bacup
30121	F135JEO	Leyland Freighter T45-180	Mountain Range	PL	1988	C53 Bamber Bridge
30122	F136JEO	Leyland Freighter T45-180	Mountain Range	PL	1988	A36 St. Annes
30123	F137JEO	Leyland Freighter T45-180	Mountain Range	PL	1988	B76 Darwen
30124	F138JEO	Leyland Freighter T45-180	Mountain Range	PL	1988	B70 Accrington
30125	L164DCW	Leyland-DAF FA60.210	Emergency One	PL	1993	C52 Fulwood
30126	L165DCW	Leyland-DAF FA60.210	Emergency One	PL	1993	A11 Lancaster
30127	L166DCW	Leyland-DAF FA60.210	Emergency One	PL	1993	A11 Lancaster
30128	M167VFV	Leyland-DAF FA60.210	LCES/Bedwas	PL	1995	A30 Blackpool
30129	M168VFV	Leyland-DAF FA60.210	LCES/Bedwas	PL	1995	A30 Blackpool

Three new Prime Movers were delivered in 1995 to Lancashire. These appliances all utilise the Leyland-DAF60-180Ti chassis and have Multilift pod handling equipment. Two of the new vehicles were allocated to Preston and one to Burnley. This view shows the Burnley machine, M701UHG, with the Foam Tender pod affixed. *Karl Sillitoe*

Lancashire County Fire Brigade have a fleet of over 70 Leyland Freighter or Leyland-DAF60 pumps supplied over a nine year period. Emergency One coachwork was introduced into the fleet on a pair of Freighter T45-180's in 1990, both of these appliances are currently at Morecambe where 30248, (H148WCK) was photographed. *Malcolm Cook*

30130	M169VFV	Leyland-DAF FA60.210	LCES/Bedwas	PL	1995	C56 Skelmersdale
30131	N170GCK	Leyland-DAF FA60.210	Emergency One	PL	1996	C51 Ormskirk
30132	N171GCK	Leyland-DAF FA60.210	Emergency One	PL	1996	Not yet allocated
30133	N172GCK	Leyland-DAF FA60.210	Emergency One	PL	1996	A31 Bispham
30134	N173GCK	Leyland-DAF FA60.210	Emergency One	PL	1996	C55 Leyland
30138	RFV934X	Bedford TK	Cheshire Fire Engineering	PL	1982	Reserve
30139	RFV935X	Bedford TK	Cheshire Fire Engineering	PL	1982	Training Centre
30146	CHG679Y	Ford Cargo 1713	Artic Towing Unit	TU	1983	Headquarters
30147	Trailer		Fire Prevention Unit	FPU	1972	Headquarters
30185	Low Loader		Semi Trailer	Tr	1972	Headquarters
30225	E114HCW	Leyland Freighter T45-180	Fulton & Wylie	PL	1988	A34 Wesham
30226	E115HCW	Leyland Freighter T45-180	Fulton & Wylie	PL	1988	A18 Garstang
30227	E116HCW	Leyland Freighter T45-180	Fulton & Wylie	PL	1988	A35 Lytham
30228	E117HCW	Leyland Freighter T45-180	Fulton & Wylie	PL	1988	B92 Padiham
30229	E118HCW	Leyland Freighter T45-180	Fulton & Wylie	PL	1988	Reserve
30230	E119HCW	Leyland Freighter T45-180	Fulton & Wylie	PL	1988	B91 Clitheroe
30231	E120HCW	Leyland Freighter T45-180	Fulton & Wylie	PL	1988	A36 St. Annes
30232	E121HCW	Leyland Freighter T45-180	Fulton & Wylie	PL	1988	Reserve
30233	E122HCW	Leyland Freighter T45-180	Fulton & Wylie	PL	1988	Reserve
30234	E123HCW	Leyland Freighter T45-180	Fulton & Wylie	PL	1988	Reserve
30235	E124HCW	Leyland Freighter T45-180	Fulton & Wylie	PL	1988	B75 Haslingdon
30236	E125HCW	Leyland Freighter T45-180	Fulton & Wylie	PL	1988	B93 Barnoldswick
30237	E126HCW	Leyland Freighter T45-180	Fulton & Wylie	PL	1988	A14 Carnforth
30238	E127HCW	Leyland Freighter T45-180	Fulton & Wylie	PL	1988	B91 Clitheroe
30239	G139JFR	Leyland Freighter T45-180	Reynolds Boughton	PL	1989	B93 Barnoldswick
30240	G140JFR	Leyland Freighter T45-180	Reynolds Boughton	PL	1989	B94 Nelson
30241	G141JFR	Leyland Freighter T45-180	Reynolds Boughton	PL	1989	A15 Silverdale
30242	G142JFR	Leyland Freighter T45-180	Reynolds Boughton	PL	1989	B74 Rawtenstall
30243	G143JFR	Leyland Freighter T45-180	Reynolds Boughton	PL	1989	C55 Leyland
30244	G144JFR	Leyland Freighter T45-180	Reynolds Boughton	PL	1989	A14 Carnforth
30245	G145JFR	Leyland Freighter T45-180	Reynolds Boughton	PL	1989	C54 Chorley
30246	G146JFR	Leyland Freighter T45-180	Reynolds Boughton	PL	1989	B96 Colne
30247	H147WCK	Leyland Freighter T45-180	Emergency One	PL	1990	A12 Morecambe
30248	H148WCK	Leyland Freighter T45-180	Emergency One	PL	1990	A12 Morecambe
30249	H149WCK	Leyland Freighter T45-180	Reynolds Boughton	PL	1990	A16 Hornby
30250	H150WCK	Leyland Freighter T45-180	Reynolds Boughton	PL	1990	A13 Bolton-Le-Sands
30251	H151WCK	Leyland Freighter T45-180	Reynolds Boughton	PL	1990	A37 South Shore, Blackpool
30252	H152WCK	Leyland Freighter T45-180	Reynolds Boughton	PL	1990	B70 Accrington
30253	H153WCK	Leyland Freighter T45-180	Reynolds Boughton	PL	1990	B73 Bacup
30254	H154WCK	Leyland Freighter T45-180	Reynolds Boughton	PL	1990	A33 Presall
30255	H155WCK	Leyland Freighter T45-180	Reynolds Boughton	PL	1990	B76 Darwen
30281	XRN512V	Land Rover Series III 88	Land Rover	L4V	1980	A11 Lancaster
30282	A791PCW	Land Rover 110	Land Rover	L4V	1984	B90 Burnley
30284	J792OCW	Land Rover Defender 110	Land Rover	L4V	1992	C54 Chorley
30285	N790GCK	Kawasaki Mule	Kawasaki	ATV	1996	C50 Preston
30286	N789LCK	Kawasaki Mule	Kawasaki	ATV	1996	C50 Preston
30323	NHG14X	Bedford TK	Cheshire Fire Engineering	PL	1981	Training Centre
30337	A705KHG	Bedford TK	HCB-Angus HSC	PL	1984	Reserve
30339	A707KHG	Bedford TK	HCB-Angus HSC	PL	1984	Training Centre
30340	A708KHG	Bedford TK	HCB-Angus HSC	PL	1984	Training Centre
30341	A709KHG	Bedford TK	HCB-Angus HSC	PL	1984	Withdrawn
30342	A710KHG	Bedford TK	HCB-Angus HSC	PL	1984	Reserve
30347	B284BFR	Bedford TK	Fulton & Wylie Fire Warrior	PL	1985	Training Centre
30348	B285BFR	Bedford TK	Fulton & Wylie Fire Warrior	PL	1985	Training Centre
30428	G500JFR	Leyland Swift LBM6T/2RS	Reeve Burgess Harrier	CU	1990	C52 Fulwood
30429	YPD101Y	Leyland Tiger TRCTL11/2R	Duple Dominant IV Exp ress	PCV	1983	Training Centre
30	P	Volvo FL10	Angloco/Bronto Skylift	ALP	1997	On Order
30	P	Volvo FL10	Angloco/Bronto Skylift	ALP	1997	On Order

Note: 30429 was formerly London Country TD1

LEICESTERSHIRE FIRE & RESCUE SERVICE

Leicestershire Fire & Rescue Service, Anstey Frith, Leicester Road, Glenfield,
Leicester LE3 8HD

JBC920V	Dodge G1313	HCB-Angus	WrL	1980	Training Centre
URY263X	Dodge G1313	HCB-Angus	WrL	1981	Training Centre
VUT311X	Dodge G1313	Angloco	ET/CU	1981	20 Loughborough
VUT312X	Dodge G1313	Angloco	ET/CU	1981	40 Leicester Southern
AFP165Y	Dennis RS133	Dennis	WrL	1982	25 Coalville
AFP166Y	Dennis RS133	Dennis	WrL	1982	31 Wigston
AFP167Y	Dennis RS133	Dennis	WrL	1982	22 Syston
ARY453Y	Dodge G13C	Carmichael	FST	1982	31 Wigston
ARY454Y	Dodge G13C	Carmichael	FST	1982	25 Coalville
BRY103Y	Dodge S66	HCB-Angus	RV	1982	38 Hinckley
BRY104Y	Dodge S66	HCB-Angus	RV	1982	33 Oakham
BUT906Y	Land Rover Series III V8	Carmichael/Hotspur	L6P	1983	Fire Safety Department
A196GUT	Dennis SS133	Dennis	WrL	1983	34 Uppingham
A197GUT	Dodge G10C	Carmichael Benson	CIU	1983	30 Leicester Central
A131KAY	Iveco-Magirus 256D14	Carmichael/Magirus DLK23-12	TL	1983	30 Leicester Central
B961NJF	Dennis SS135	Dennis	WrL	1984	34 Uppingham
B962NJF	Dennis SS135	Dennis	WrL	1984	33 Oakham
C675UUT	Dodge G13C	Saxon Sanbec	WrL	1985	26 Ashby-de-la-Zouch
C676UUT	Dodge G13C	Saxon Sanbec	WrL	1985	21 Melton Mowbray
C677UUT	Dodge G13C	Saxon Sanbec	WrL	1985	21 Melton Mowbray
C678UUT	Dodge G13C	Saxon Sanbec	WrL	1985	27 Moira
C679UUT	Dodge G13C	Saxon Sanbec	WrL	1985	32 Billesdon
C503WAY	Dodge S66	Leicester Carriage	CaV	1986	30 Leicester Central

**The Leicestershire Fire & Rescue Service have supported the local coachbuilder, Leicester Carriage
Company, with orders for two appliances. The coachwork for the Incident Control Unit bears a slight
resemblance to the Willowbrook Warrier bus of the same era. Now relocated at the new showpiece
station at Leicester Southern, G580YJU illustrates its unusual design.** *Clive Shearman*

The Leicestershire Fire and Rescue Service joined the increasing number of Brigades operating Dennis Sabre appliances when, in 1995, four TSD233s arrived. Leicester Central received N775OJF fitted out as a Water Tender Ladder. *Tony Wilson*

D874YFP	Shelvoke & Drewry WY	Carmichael/Bronto Skylift 27.3	HP	1986	30 Leicester Central
D946BJF	Dennis SS135	Saxon Sanbec	WrL	1986	36 Market Harborough
D947BJF	Dennis SS135	Saxon Sanbec	WrL	1986	37 Lutterworth
D948BJF	Dennis SS135	Saxon Sanbec	WrL	1986	28 Shepshed
F378PJU	Dennis SS135	Saxon Sanbec	WrL	1988	36 Market Harborough
F379PJU	Dennis SS135	Saxon Sanbec	WrL	1988	39 Market Bosworth
G580YJU	Leyland-DAF Roadrunner	Leicester Carriage	ICU	1989	40 Leicester Southern
G581YJU	Dennis SS237	Excalibur CBK	WrL	1989	38 Hinckley
G582YJU	Dennis SS237	Excalibur CBK	WrL	1989	35 Kibworth
G583YJU	Dennis SS237	Excalibur CBK	WrL	1989	Reserve
G584YJU	Dennis SS237	Excalibur CBK	WrL	1989	Reserve
H857ERY	Dennis SS237	Excalibur CBK	WrL	1990	Reserve
H858ERY	Dennis SS237	Excalibur CBK	WrL	1990	25 Coalville
H859ERY	Dennis SS237	Excalibur CBK	WrL	1990	33 Oakham
H860ERY	Dennis SS237	Excalibur CBK	WrL	1990	20 Loughborough
H304GFP	Land Rover Defender 110	Land Rover	PCV	1990	25 Coalville
J671MNR	Land Rover Defender 110	Land Rover/Sturgess	PCV	1991	33 Oakham
J672MNR	Land Rover Defender 110	Land Rover/Sturgess	PCV	1991	38 Hinckley
J513MJF	Dennis SS237	Excalibur CBK	WrL	1991	Reserve
J514MJF	Dennis SS237	Excalibur CBK	WrL	1991	24 Leicester Western
J515MJF	Dennis SS237	Excalibur CBK	WrL	1991	38 Hinckley
J516MJF	Dennis SS237	Excalibur CBK	WrL	1992	20 Loughborough
K787TNR	Dennis Rapier TF202	Excalibur CBK	WrL	1992	Training Centre
K788TNR	Dennis SS237	Excalibur CBK	WrL	1992	24 Leicester Western
K789TNR	Dennis SS237	Excalibur CBK	WrL	1992	31 Wigston
K894UBC	Land Rover Defender 110	Land Rover	PCV	1992	20 Loughborough
L404DJF	Dennis SS237	Excalibur CBK	WrL	1993	23 Leicester Eastern
L405DJF	Dennis SS237	Excalibur CBK	WrL	1993	30 Leicester Central
N773OJF	Dennis Sabre TSD233	John Dennis	WrL	1995	20 Loughborough
N774OJF	Dennis Sabre TSD233	John Dennis	WrL	1995	40 Leicester Southern
N775OJF	Dennis Sabre TSD233	John Dennis	WrL	1995	30 Leicester Central
N776OJF	Dennis Sabre TSD233	John Dennis	WrL	1995	23 Leicester Eastern
N665SFP	Dennis Sabre TSD233	John Dennis	WrL	1995	24 Leicester Western

LINCOLNSHIRE FIRE BRIGADE

Lincolnshire County Fire Brigade, South Park Avenue, Lincoln LN5 8EL

FW9545	Leyland	Leyland	PE	1937	B1 Skegness (Preserved)
JL7506	Leyland	Leyland	WrT	1938	C1 Boston (Preserved)
JL5726	Leyland	Leyland	WrT	1940	D1 Grantham (Preserved)
TTL344R	Bedford TK	HCB-Angus	WrT	1977	C Reserve - Mothballed
XRO614S	Dodge K1613	Eagle/Angloco	WrC	1977	C2 Spalding
LFW596V	Bedford TK	HCB-Angus HSC	WrL	1980	Reserve
LFW597V	Bedford TK	HCB-Angus HSC	WrL	1980	Mothballed
LFW598V	Bedford TK	HCB-Angus HSC	WrL	1980	On loan to industry
LFW599V	Bedford TK	HCB-Angus HSC	WrT	1980	On loan to industry
Pod	Transliner	Canteen Van	CaV	1980	C1 Boston
RVL110W	Bedford TKG	HCB-Angus HSC	WrT	1980	D6 Corby Glen
RVL111W	Bedford TKG	HCB-Angus HSC	WrL	1980	A1 Lincoln
UFW124W	Bedford TKG	HCB-Angus HSC	WrT	1981	Mothballed
UFW125W	Bedford TKG	HCB-Angus HSC	WrT	1981	Mothballed
AFW122X	Bedford TKG	HCB-Angus HSC	WrL	1981	A Reserve
AFW123X	Bedford TKG	HCB-Angus HSC	WrL	1981	D Reserve
AFW124X	Bedford TKG	HCB-Angus HSC	WrL	1981	B Reserve
AFW125X	Bedford TKG	HCB-Angus HSC	WrL	1981	C Reserve
Pod	Transliner	Exhibition Unit	ExU	1982	A1 Lincoln
Pod	Charles Warner	Incident Control Unit	ICU	1982	C1 Boston
HFE205Y	Bedford TK1260	Warner Woodcock	RT	1982	A1 Lincoln
HFE221Y	Bedford TKG	HCB-Angus HSC(198.)	WrT	1983	D4 Bourne
HFE222Y	Bedford TKG	Merryweather	WrT	1983	D3 Billinghay
Pod	Charles Warner	Equipment unit	EqU	1983	D1 Grantham
Pod	Transliner	Exhibition unit	ExU	1983	C1 Boston
Pod	Charles Warner	Flat bed unit	FBL	1983	A1 Lincoln
Pod	Charles Warner	Flat bed unit	FBL	1983	C1 Boston
Pod	Charles Warner	Flat bed unit	FBL	1983	D1 Grantham
A702PVL	Bedford TKG	HCB-Angus HSC	WrT	1983	B7 North Somercotes
A703PVL	Bedford TKG	HCB-Angus HSC	WrT	1983	D8 Metheringham

Lincolnshire's fleet of nine Rescue Tenders was recently reduced to six following the delivery of four new Rescue Pumps early in 1996. Newest of the remaining Rescue Tenders is H826DVL, now based at Stamford. This is one of a trio of Saxon Sanbec-bodied Volvo FL6s, the other example having a Bedford chassis. *Andrew Fenton*

A714PVL	Bedford TL1260	Powell Dyffryn Rolonof	PM	1983	C1 Boston
Pod	Transliner	Canteen Van	CaV	1984	A1 Lincoln
Pod	Transliner	Exhibition Unit	ExU	1984	B1 Skegness
B604AVL	Bedford TKG	HCB-Angus HSC	WrT	1984	A9 Wragby
B605AVL	Bedford TKG	HCB-Angus HSC	WrT	1984	B3 Binbrook
B606AVL	Bedford TKG	HCB-Angus HSC	WrT	1984	C5 Holbeach
B350DFW	Bedford TL1260	Powell Dyffryn Rolonof	PM	1985	A1 Lincoln
Pod	Transliner	Equipment Unit	EqU	1985	C1 Boston
Pod	Charles Warner	Flat bed unit	FBL	1985	B1 Skegness
C562LVL	Bedford TKG	HCB-Angus HSC	WrL	1986	C2 Spalding
C563LVL	Bedford TKG	HCB-Angus HSC	WrT	1986	B8 Spilsby
C564LVL	Bedford TKG	HCB-Angus HSC	WrT	1986	A8 Waddington
C570LVL	Bedford TL750	Saxon Sanbec	RT	1986	C Reserve
Pod	Transliner	Equipment Unit	EqU	1986	B1 Skegness
Pod	Bazoo/Argocat	Coastal Pollution Unit	CPU	1986	C1 Boston
D94TTL	Bedford TL	HCB-Angus HSC	WrT	1986	B9 Wainfleet
D96TTL	Bedford TL	HCB-Angus HSC	WrL	1986	B6 Mablethorpe
D97TTL	Bedford TL	HCB-Angus HSC	WrT	1986	D7 Market Deeping
D95TTL	Bedford TL	HCB-Angus HSC	WrT	1987	C3 Crowland
D118WTL	Bedford TL750	HCB-Angus	RT	1987	D9 Sleaford
D799XTL	Bedford TL	HCB-Angus HSC	WrT	1987	D5 Brant Broughton
D801XTL	Bedford TL	HCB-Angus HSC	WrT	1987	A7 Saxilby
E906AVL	Bedford TL	HCB-Angus HSC	WrL	1987	A2 Gainsborough
E213DFE	Volvo FL6.14	HCB-Angus	WrT	1988	A6 North Hykeham
E214DFE	Volvo FL6.14	HCB-Angus	WrL	1988	A1 Lincoln
E215DFE	Volvo FL6.14	HCB-Angus	WrL	1988	D9 Sleaford
E216DFE	Volvo FL6.14	HCB-Angus	WrL	1988	D1 Grantham
F576HTL	Renault-Dodge G16C	Saxon Sanbec/Simon SS263	HP	1988	A1 Lincoln
F591HTL	Volvo FL6.14	HCB-Angus	WrT	1989	B2 Alford
F592HTL	Volvo FL6.14	HCB-Angus	WrT	1989	B5 Louth
F593HTL	Volvo FL6.14	HCB-Angus	WrL	1989	D10 Stamford
F594HTL	Volvo FL6.14	HCB-Angus	WrT	1989	D10 Stamford
G662TFW	Renault-Dodge G16C	Saxon Sanbec/Simon SS263	HP	1989	C1 Boston
G432VFE	Volvo FL6.14	Saxon Sanbec	WrL	1989	B1 Skegness
G433VFE	Volvo FL6.14	Saxon Sanbec	WrL	1989	B4 Horncastle
G434VFE	Volvo FL6.14	Saxon Sanbec	WrL	1989	C9 Woodhall Spa
G435VFE	Volvo FL6.14	Saxon Sanbec	RT	1989	D1 Grantham
Pod	Transliner	Equipment Unit	EQU	1989	A1 Lincoln
H348CVL	Volvo FL6.14	Saxon Sanbec	WrL	1990	C8 Long Sutton
H349CVL	Volvo FL6.14	Saxon Sanbec	WrT	1990	D2 Billingborough
H355CVL	Volvo FL6.14	Saxon Sanbec	WrT	1990	C7 Leverton
H356CVL	Volvo FL6.14	Saxon Sanbec	WrT	1990	A3 Bardney
H357CVL	Volvo FL6.14	Lincs FB	DTV	1990	Driving School
H826DVL	Volvo FL6.14	Saxon Sanbec	RT	1991	C1 Boston
Pod	Charles Warner	Incident Control Unit	ICU	1991	A1 Lincoln
Pod	Charles Warner	BA Unit	BAU	1991	A1 Lincoln
J632HFW	Volvo FL6.14	Excalibur CBK	WrT	1991	C6 Kirton
J633HFW	Volvo FL6.14	Excalibur CBK	WrT	1991	C4 Donington
J634HFW	Volvo FL6.14	Excalibur CBK	WrL	1991	A5 Market Rasen
J635HFW	Volvo FL6.14	Excalibur CBK	WrT	1991	A4 Caistor
J641HFW	Volvo FL6.14	Saxon Sanbec	RT	1992	B1 Skegness
Pod	Charles Warner	Chemical Incident Unit	CIU	1992	D1 Grantham
K392PVL	Volvo FL6.18	Bedwas/Simon Snorkel SS263	HP	1993	B1 Skegness
K908RFW	Volvo FL6.14	Excalibur CBK	WrL	1993	D1 Grantham
K909RFW	Volvo FL6.14	Excalibur CBK	WrL	1993	A1 Lincoln
K910RFW	Volvo FL6.14	Excalibur CBK	WrL	1993	A1 Lincoln
L596YVL	Volvo FL6.14	Excalibur CBK	WrL	1994	C1 Boston
L597YVL	Volvo FL6.14	Excalibur CBK	WrL	1994	C1 Boston
L598YVL	Volvo FL6.14	Excalibur CBK	WrL	1994	B1 Skegness
L729VVL	Land Rover Defender	Land Rover	L4V	1993	C1 Boston
L730VVL	Land Rover Defender	Land Rover	L4V	1993	D1 Grantham
L732VVL	Land Rover Defender	Land Rover	L4V	1993	B1 Skegness
M671GVL	Volvo FL6.14	Powell Duffryn Rolonof	PM	1995	D1 Grantham
M914JFE	Volvo FL6.14	Powell Duffryn Rolonof	PM	1995	B1 Skegness
N905JFE	Volvo FL6.14	Excalibur CBK	WrL/R	1996	A2 Gainsborough
N906JFE	Volvo FL6.14	Excalibur CBK	WrL/R	1996	B5 Louth
N907JFE	Volvo FL6.14	Excalibur CBK	WrL/R	1996	D9 Sleaford
N908JFE	Volvo FL6.14	Excalibur CBK	WrL/R	1996	C2 Spalding
N947KTL	Land Rover Defender	Land Rover	L4V	1995	A1 Lincoln

Note: H357CVL was rebuilt from a WrL in 1996; XRO614S was aquired from Bedfordshire FRS in 1992
On Order: - 4 Volvo FL6.14/Excalibur CBK and 1 Dennis Sabre S411/John Dennis.

LONDON FIRE BRIGADE

London Fire Brigade, 8 Albert Embankment, Lambeth, SE1 7SD

ALP1-6

Volvo FL10 8x4 — Angloco/Bronto 33-2Ti — ALP — 1990-91

1	H901XYF	A28 Dowgate	3	H903XYF	H34 Wimbledon	5	H905XYF	E31 Forest Hill
2	H902XYF	A33 Tottenham	4	H904XYF	G30 Wembley	6	H906XYF	F45 Plaistow

CSU 1-6

Mercedes-Benz 917AF — Locomotors — CSU — 1991

1	H861XYF	Training/Reserve	3	H863XYF	E20 Lewisham SC	5	H865XYF	F20 Stratford EC
2	H862XYF	G20 Wembley WC	4	H864XYF	H31 Croydon	6	H866XYF	West Reserve

DPL722 NYV722Y — Dennis SS131 — Dennis — WrT — 1983 — West Training

DPL 736-801

Dennis SS131 — Dennis — RP — 1984

736	A736SUL	South Reserve	756	A756SUL	South Training	783	A783TYO	West Reserve
737	A737SUL	Training Centre	757	A757SUL	South Reserve	785	A785TYO	West Reserve
738	A738SUL	South Reserve	760	A760SUL	Training Centre	786	A786TYO	Withdrawn
739	A739SUL	South Reserve	762	A762SUL	Withdrawn	787	A787TYO	West Reserve
740	A740SUL	East Reserve	763	A763SUL	South Reserve	788	A788TYO	West Reserve
741	A741SUL	Training Centre	764	A764SUL	F35 Woodford	791	A791TYO	East Reserve
743	A743SUL	South Reserve	765	A765SUL	South Reserve	792	A792TYO	East Reserve
744	A744SUL	South Reserve	767	A767SUL	South Reserve	793	A793TYO	Withdrawn
745	A745SUL	South Reserve	769	A769SUL	West Reserve	795	A795TYO	East Reserve
746	A746SUL	Withdrawn	771	A771SUL	South Reserve	796	A796TYO	East Reserve
747	A747SUL	South Reserve	773	A773SUL	Training Centre	797	A797TYO	South Reserve
748	A748SUL	East Reserve	774	A774SUL	East Reserve	798	A798TYO	East Reserve
749	A749SUL	Training Centre	776	A776SUL	East Reserve	799	A799TYO	West Reserve
751	A751SUL	South Reserve	778	A778TYO	West Reserve	800	A800TYO	East Reserve
752	A752SUL	East Reserve	779	A779TYO	Withdrawn	801	A801TYO	West Reserve
754	A754SUL	F46 Silvertown	781	A781TYO	West Reserve			

DPL 803-824

Renault-Dodge G13 — Locomotors — RP — 1987

803	D803FYM	East Reserve	811	D811FYM	E23 East Greenwich	819	D819FYM	Training Centre
804	D804FYM	Driving School	812	D812FYM	H41 Kingston	820	D820FYM	H21 Clapham
805	D805FYM	G24 Southall	813	D813FYM	East Reserve	821	D821FYM	East Reserve
806	D806FYM	A37 Barnet	815	D815FYM	West Reserve	822	D822FYM	East Reserve
808	D808FYM	West Training	817	D817FYM	South Training	823	D823FYM	South Reserve
809	D809FYM	West Reserve	818	D818FYM	East Reserve	824	D824FYM	Driving School
810	D810FYM	West Reserve						

Some 108 Dennis SS131 pumps were delivered to the London Fire Brigade during the latter years of GLC control. Deliveries of new Volvo FL6 pumps have almost eliminated these from front line service and all bar two are now used for training or reserves. Seen in September 1996 at Edmonton is DPL798, A798TYO.
Keith Grimes

DPL 825-830 — Renault-Dodge G13 — Saxon Sanbec — RP 1987

No.	Reg	Location	No.	Reg	Location	No.	Reg	Location
825	D825FYM	West Reserve	828	D828FYM	G35 Fulham	830	D830FYM	West Reserve
827	D827FYM	E34 Edmonton	829	D829FYM	East Reserve			

DPL 831-834 — Renault-Dodge G13 — Locomotors — RP 1987

No.	Reg	Location	No.	Reg	Location	No.	Reg	Location
831	D831FYM	F33 Whitechapel	833	D833FYM	Training Centre	834	D834FYM	E27 Erith
832	D832FYM	South Reserve						

DPL 835-866 — Renault-Dodge G13 — Saxon Sanbec — RP 1988

No.	Reg	Location	No.	Reg	Location	No.	Reg	Location
835	E835JYV	E40 Sidcup	846	E846JYV	E36 Deptford	857	E857JYV	E33 Wandsworth
836	E836JYV	E39 Bromley	847	E847JYV	H24 Brixton	858	E858JYV	H43 Twickenham
837	E837JYV	H34 Wimbledon	848	E848JYV	H26 Addington	859	E859JYV	G27 North Kensington
838	E838JYV	E33 Southwark	849	E849JYV	G25 Ealing	860	E860JYV	Training Centre
839	E839JYV	G23 Hillingdon	850	E850JYV	G33 Kensington	861	E861JYV	A21 Paddington
840	E840JYV	G26 Acton	851	E851JYV	F45 Plaistow	862	E862JYV	A43 Kentish Town
841	E841JYV	A32 Hornsey	852	E852JYV	A25 Westminster	863	E863JYV	F31 Kingsland
842	E842JYV	West Reserve	853	E853JYV	F24 Shoreditch	864	E864JYV	F39 Hornchurch
843	E843JYV	F28 Homerton	854	E854JYV	F22 Poplar	865	E865JYV	A41 West Hampstead
844	E844JYV	F26 Bethnal Green	855	E855JYV	E31 Forest Hill	866	E866JYV	G38 Heston
845	E845JYV	South Training	856	E856JYV	West Training			

DPL 867-900 — Volvo FL6.14 — Saxon Sanbec — RP 1990-91

No.	Reg	Location	No.	Reg	Location	No.	Reg	Location
867	H867XYF	F30 Leytonstone	879	H879XYF	F34 Chingford	890	H890XYF	A41 West Hampstead
868	H868XYF	E37 Peckham	880	H880XYF	A35 Enfield	891	H891XYF	G32 Ruislip
869	H869XYF	F41 Dagenham	881	H881XYF	G40 Hayes	892	H892XYF	H32 Norbury
870	H870XYF	A31 Holloway	882	H882XYF	H32 Norbury	893	H893XYF	E24 Woolwich
871	H871XYF	G37 Chiswick	883	H883XYF	E42 Biggin Hill	894	H894XYF	F35 Woodford
872	H872XYF	H22 Lambeth	884	H884XYF	F43 Barking	895	H895XYF	A39 Finchley
873	H873XYF	E22 Greenwich	885	H885XYF	A43 Kentish Town	896	H896XYF	H43 Twickenham
874	H874XYF	F27 Bow	886	H886XYF	G39 Feltham	897	H897XYF	E30 Eltham
875	H875XYF	A35 Enfield	887	H887XYF	H37 Wallington	898	H898XYF	F32 Stoke Newington
876	H876XYF	G21 Harrow	888	H859XYF	E35 Old Kent Road	899	H899XYF	A26 Knightsbridge
877	H877XYF	H40 New Malden	889	H889XYF	F36 Walthamstow	900	H860XYF	H26 Addington
878	H878XYF	E27 Erith						

DPL 901-924 — Volvo FL6.14 — Saxon Sanbec — RP 1992

No.	Reg	Location	No.	Reg	Location	No.	Reg	Location
901	J901CYP	A24 Soho	909	J909CYP	H31 Croydon	917	J917CYP	F22 Poplar
902	J902CYP	F44 East Ham	910	J910CYP	G29 Park Royal	918	J918CYP	E34 Dockhead
903	J903CYP	E23 East Greenwich	911	J911CYP	A23 Euston	919	J919CYP	H39 Surbiton
904	J904CYP	H29 Purley	912	J912CYP	F38 Romford	920	J920CYP	G34 Chelsea
905	J905CYP	G22 Stanmore	913	J913CYP	E28 Bexley	921	J921CYP	A26 Knightsbridge
906	J906CYP	A22 Manchester Sq	914	J914CYP	H31 Croydon	922	J922CYP	F33 Whitechapel
907	J907CYP	A30 Islington	915	J915CYP	G31 Northolt	923	J923CYP	E33 Southwark
908	J908CYP	E31 Forest Hill	916	J916CYP	A40 Hendon	924	J924CYP	G36 Hammersmith

In 1988, London Fire Brigade took delivery of a batch of Renault-Dodge G13 pumps with Saxon Sanbec bodies. Representing the batch is DPL855, E855JYV, seen in Shootrs Hill, London.
Gerald Mead

When the Light Rescue Units were converted to Command Support Units, the Heavy Rescue Units were redesignated as Fire Rescue Units (FRU). FRU1 (H851XYF) is based at Lewisham in the southern area command and was photographed while in attendance at a collapse of scaffolding in High Street, Sidcup on 12th June 1996. All six FRUs have Carmichael bodywork on Volvo FL6.14 chassis. *Glyn Matthews*

DPL 925-949   Volvo FL6.14   Saxon Sanbec   RP   1993

925	K925EYH	A27 Clerkenwell	934	K934EYH	H27 Battersea	942	K942EYH	G33 Kensington
926	K926EYH	F21 Stratford	935	K935EYH	A33 Tottenham	943	K943EYH	E36 Deptford
927	K927EYH	G28 Willesden	936	K936EYH	F30 Leytonstone	944	K944EYH	H27 Battersea
928	K928EYH	E35 Old Kent Road	937	K937EYH	G27 Nth Kensington	945	K945EYH	A42 Belsize
929	K929EYH	H36 Mitcham	938	K938EYH	E25 Plumstead	946	K946EYH	F46 Silvertown
930	K930EYH	A29 Barbican	939	K939EYH	H28 Woodside	947	K947EYH	G25 Ealing
931	K931EYH	F31 Kingsland	940	K940EYH	A37 Barnet	948	K948EYH	E39 Bromley
932	K932EYH	G23 Hillingdon	941	K941EYH	F42 Ilford	949	K949EYH	H28 Woodside
933	K933EYH	E21 Lewisham						

DPL 950-971   Volvo FL6.14   Saxon Sanbec   RP   1994

950	L950GYK	A31 Holloway	958	L958GYK	E38 New Cross	965	L965GYK	A36 Southgate
951	L951GYK	F25 Shadwell	959	L959GYK	H38 Sutton	966	L966GYK	F45 Plaistow
952	L952GYK	G24 Southall	960	L960GYK	A28 Dowgate	967	L967GYK	G38 Heston
953	L953GYK	E43 Beckenham	961	L961GYK	F28 Homerton	968	L968GYK	E32 Downham
954	L954GYK	H25 West Norwood	962	L962GYK	G35 Fulham	969	L969GYK	A34 Edmonton
955	L955GYK	A25 Westminster	963	L963GYK	E40 Sidcup	970	L970GYK	F37 Hainault
956	L956GYK	F23 Millwall	964	L964GYK	H42 Richmond	971	L971GYK	F39 Hornchurch
957	L957GYK	G30 Wembley						

DPL 972-994   Volvo FL6.14   Saxon Sanbec   RP   1995

972	M972LYP	A21 Paddington	980	M980LYP	G36 Hammersmith	988	M988LYP	F27 Bow
973	M973LYP	F26 Bethnal Green	981	M981LYP	H35 Tooting	989	M989LYP	G28 Willesden
974	M974LYP	G39 Feltham	982	M982LYP	West Training	990	M990LYP	A39 Finchley
975	M975LYP	E26 Shooters Hill	983	M983LYP	F21 Stratford	991	M991LYP	F42 Ilford
976	M976LYP	H35 Tooting	984	M984LYP	G26 Acton	992	M992LYP	G34 Chelsea
977	M977LYP	A32 Hornsey	985	M985LYP	E29 Lee Green	993	M993LYP	A38 Mill Hill
978	M978LYP	F24 Shoreditch	986	M986LYP	H33 Wandsworth	994	M994LYP	F32 Stoke Newington
979	M979LYP	E41 Orpington	987	M987LYP	A30 Islington			

One of a pair of King & Taylor-bodied HDCs is HDC29, J929CYP. Pictured in 1996 at Heathrow airport when fitted with a **Bulk Foam unit.** *Gerald Mead*

DPL 995-1016   Volvo FL6.14   Saxon Sanbec   RP   1996

995	N21OHX	F41 Dagenham	1003	N31OHX	H21 Clapham	1010	N39OHX	G40 Hayes
996	N23OHX	F43 Barking	1004	N32OHX	H41 Kingston	1011	N41OHX	G37 Chiswick
997	N24OHX	A27 Clerkenwell	1005	N34OHX	E25 Plumstead	1012	N42OHX	A40 Hendon
998	N25OHX	F38 Romford	1006	N35OHX	E22 Greenwich	1013	N43OHX	G30 Wembley
999	N26OHX	F40 Wennington	1007	N36OHX	E34 Dockhead	1014	N45OHX	A24 Soho
1000	N27OHX	F29 Leyton	1008	N37OHX	H34 Wimbledon	1015	N46OHX	A23 Euston
1001	N28OHX	A33 Tottenham	1009	N38OHX	H24 Brixton	1016	N47OHX	A22 Manchester Square
1002	N29OHX	E37 Peckham						

DPL 1017-1034   Volvo FL6.14   Saxon Sanbec   RP   1997

1017	P467RHV		1023	P473RHV		1029	P479RHV
1018	P468RHV		1024	P474RHV		1030	P480RHV
1019	P469RHV		1025	P475RHV		1031	P481RHV
1020	P470RHV		1026	P476RHV		1032	P482RHV
1021	P471RHV		1027	P477RHV		1033	P483RHV
1022	P472RHV		1028	P478RHV		1034	P484RHV

FRU 1-6   Volvo FL6.14   Carmichael   FRU   1991-92

1	H851XYF	E21 Lewisham	3	H853XYF	A23 Euston	5	J935CYP	F44 East Ham
2	H852XYF	H31 Croydon	4	H854XYF	G38 Heston	6	J936CYP	A39 Finchley

GPL 1-5   Iveco TurboDaily 59-12   J S Keams   GPL   1996

1	N994OHV	F20 Stratford EC	3	N996OHV	E20 Lewisham SC	5	N998OHV	Training Centre
2	N995OHV	F20 Stratford EC	4	N997OHV	G20 Wembley WC			

Code	Reg	Chassis	Body	Fleet	Year	Location
HDC12	A812TYO	Dodge RG13C	Fruehauf	HDC	1984	Ruislip Works

HDC 16-20
Renault-Dodge RG13 · Fruehauf · HDC 1987

No	Reg	Location	No	Reg	Location	No	Reg	Location
16	D866FYM	West Reserve	18	D868FYM	East Reserve	20	D870FYM	South Training
17	D867FYM	West Reserve	19	D869FYM	South Reserve			

Code	Reg	Chassis	Body	Year	Location
HDC21	F641OYL	Renault-Dodge RG13	Fruehauf	1988	South Reserve
HDC22	G922TYN	Volvo FL6.14	Locomotors	1989	South Reserve

HDC 23-27
Volvo FL6.14 · Arlington · HDC 1990

No	Reg	Location	No	Reg	Location	No	Reg	Location
23	G923TYN	West Reserve	25	G925TYN	F43 Barking	27	G927TYN	South Reserve
24	G924TYN	East Reserve	26	G926TYN	South Reserve			

Code	Reg	Chassis	Body		Year	Location
HDC28	J928CYP	Volvo FL6.14	King & Taylor		1992	A34 Edmonton
HDC29	J929CYP	Volvo FL6.14	King & Taylor		1992	West Reserve
HDC30	K960EYH	Volvo FL6.14	Locomotors		1993	A34 Edmonton
HDC31	K961EYH	Volvo FL6.14	Locomotors		1993	E20 Lewisham SC
HDC32	K962EYH	Volvo FL6.14	Locomotors		1993	H31 Croydon
HDC33	L21GYK	Volvo FL6.14	Hilbrow		1994	G20 Wembley WC
HDC34	L23GYK	Volvo FL6.14	Hilbrow		1994	F43 Barking
HDC35	L24GYK	Volvo FL6.14	Hilbrow		1994	G40 Hayes
HDC36	M996LYP	Volvo FL6.14	Saxon Sanbec		1995	E43 Beckenham
HDC37	M997LYP	Volvo FL6.14	Saxon Sanbec		1995	G40 Hayes
HDC38	M998LYP	Volvo FL6.14	Saxon Sanbec		1995	F20 Stratford EC
HDC39	M970LYP	Volvo FL6.14	Saxon Sanbec		1995	H38 Sutton
HDC40	P465RHV	Volvo FL6.14	Saxon Sanbec		1996	H38 Sutton
HDC41	P466RHV	Volvo FL6.14	Saxon Sanbec		1996	E43 Beckenham
HP3	GYW663W	Shelvoke & Drewry WY	CFE/Saxon /Simon Snorkel SS220	HP	1980	East Reserve
HP11	KUV696X	Shelvoke & Drewry WY	CFE/Simon Snorkel SS220	HP	1981	West Reserve

HP 13-15
Dennis F125 · Dennis/Simon Snorkel SS220 · 1982

No	Reg	Location	No	Reg	Location	No	Reg	Location
13	NYV773Y	South Reserve	14	NYV774Y	West Reserve	15	NYV775Y	South Reserve

Code	Reg	Chassis	Body		Year	Location
HP16	A786SUL	Dodge G16	Saxon/Simon Snorkel SS220	HP	1984	West Reserve
HP17	A787SUL	Dodge G16	Saxon/Simon Snorkel SS220	HP	1984	South Reserve
HP18	K959EYH	Volvo FL6.18	Saxon/Simon Snorkel SS220(1980)	HP	1993	South Reserve
HP19	N991OHV	Volvo FL6.18	Angloco/Simon Snorkel SS220(1980)	HP	1996	H43 Twickenham
HP20	N992OHV	Volvo FL6.18	Saxon/Simon Snorkel SS220(1980)	HP	1996	E22 Greenwich
HP21	N993OHV	Volvo FL6.18	Saxon/Simon Snorkel SS220(1980)	HP	1996	H32 Norbury
HP22	P462RHV	Volvo FL6.18	Saxon/Simon Snorkel SS220(1980)	HP	1997	
HP23	P463RHV	Volvo FL6.18	Saxon/Simon Snorkel SS220(1980)	HP	1997	

MDC 1-31
Renault-Dodge S56 · Rawson · 1988

No	Reg	Location	No	Reg	Location	No	Reg	Location
1	E901JYV	G20 Wembley WC	12	E912JYV	Training Centre	22	E922JYV	F20 Stratford EC
2	E902JYV	E20 Lewisham SC	13	E913JYV	E20 Lewisham SC	23	E923JYV	F20 Stratford EC
3	E903JYV	E20 Lewisham SC	14	E914JYV	East Reserve	24	E924JYV	E20 Lewisham SC
4	E904JYV	Withdrawn	15	E915JYV	E20 Lewisham SC	25	E924JYV	E20 Lewisham SC
5	E905JYV	E20 Lewisham SC	16	E916JYV	G26 Acton	26	E926JYV	F20 Stratford EC
6	E906JYV	Training Centre	17	E917JYV	G20 Wembley WC	27	E927JYV	G20 Wembley WC
7	E907JYV	Training Centre	18	E918JYV	East Reserve	28	E928JYV	G20 Wembley WC
9	E909JYV	H21 Clapham	19	E919JYV	F20 Stratford EC	29	E929JYV	F20 Stratford EC
10	E910JYV	E20 Lewisham SC	20	E920JYV	E20 Lewisham SC	30	E930JYV	F20 Stratford EC
11	E911JYV	F20 Stratford EC	21	E921JYV	E20 Lewisham SC	31	E831JYV	Ruislip Works

MDC 32-49
Renault-Dodge S56 · King & Taylor · 1990

No	Reg	Location	No	Reg	Location	No	Reg	Location
32	G232UUW	G20 Wembley WC	38	G238UUW	E20 Lewisham SC	45	G245UUW	F44 East Ham
33	G233UUW	E20 Lewisham SC	39	G239UUW	E20 Lewisham SC	46	G246UUW	A27 Clerkenwell
34	G234UUW	West Training	41	G241UUW	Training Centre	47	G247UUW	E20 Lewisham SC
35	G235UUW	G20 Wembley WC	42	G242UUW	Training Centre	48	G248UUW	E20 Lewisham SC
36	G236UUW	F20 Stratford EC	43	G243UUW	G20 Wembley WC	49	G249UUW	G20 Wembley WC
37	G237UUW	E38 New Cross	44	G244UUW	E20 Lewisham SC			

Eight Dodge G16 with Carmichael bodywork and Magirus ladders are in the London Fire Brigades fleet. TL43 (A793SUL) is a reserve appliance that was photographed while negotiating traffic when answering a shout from Soho fire station. The aerial appliance at Soho is generally the busiest machine in London, by virtue of an aerial appliance being on the pre-determined attendance to most addresses in the West End. *Colin Lloyd*

MRS7	N841OYN	Vauhall Astravan 1.7D	Vauxhall	1996	Operations Dept
MRS8	N842OYN	Vauhall Astravan 1.7D	Vauxhall	1996	Operations Dept
MRS9	N843OYN	Vauhall Brava 2.5D	Vauxhall	1996	Personnel Section
P272	ECG272	Dennis Big4	Dennis Braidwood	1941	Ruislip Works (Preserved)
P446	JLT446K	Dennis F108	Dennis	1972	Ruislip Works (Preserved)
STP1	D826FYM	Renault-Dodge G13	Saxon Sanbec/Astatic Skidmaster	1987	Training Centre
STP2	D814FYM	Renault-Dodge G13	Locomotors	1987	Training Centre
STP3	D816FYM	Renault-Dodge G13	Locomotors	1987	Training Centre
TL19	CBY1	AEC Regent V	Merryweather	1963	Training Centre (Preserved/Ceremonial)
TL40	NYV790Y	Dodge G16	Carmichael/Magirus DL30U	1983	A21 Paddington
TL41	NYV791Y	Dodge G16	Carmichael/Magirus DL30U	1983	West Reserve

TL 42-44		Dodge G16		Carmichael/Magirus DL30E		1983		
42	A792SUL	A30 Islington	43	A793SUL	West Reserve	44	A794SUL	East Reserve

TL 45-47		Dodge G16		Carmichael/Magirus DL30U		1984		
45	B545WYO	South Reserve	46	B546WYO	E35 Old Kent Road	47	B547WYO	H21 Clapham

TL 48-50		Dennis F127		John Dennis/Camiva EPA30		1988		
48	E868JYV	F29 Leyton	49	E869JYV	A24 Soho	50	E870JYV	G34 Chelsea

FIREBOATS				
Firehawk		Watercraft Fireboat (18.74 tons)	1976	H23 Lambeth River (Reserve)
The London Pheonix		Fairey Marine Catamaran (40 tons)	1985	H23 Lambeth River
Tender 1		Dell Quay Rigid Raider	1985	H23 Lambeth River
Tender 2		Lifeguard M390 Inflatable	1989	H23 Lambeth River
Tender 3		Avon Inflatable	1986	H23 Lambeth River

Four hydraulic platform appliances of London Fire Brigade have been moved onto new Volvo FL6 chassis from Shelvoke & Drewry products built in 1980. HP19, N991OHV, is the only example to feature Angloco bodywork. This view shows the vehicle shortly after delivery. *Graham Brown*

DEMOUNTABLE BODIES FOR HEAVY CHASSIS UNITS (HDC'S)

BATB 1-6
Penman/Ray Smith (B.A.Training Body)   1993

1	Training Centre	3	South Training	5	West Training
2	South Training	4	West Training	6	East Training

BATB 7-12
Penman (B.A.Training Body)   1994

7	South Training	9	West Training	11	East Training
8	South Training	10	West Training	12	Training Centre

BFB 1-4
Reynolds Boughton (Bulk Foam Unit Body)   1992

1	G40 Hayes	3	E43 Beckenham	4	Reserve
2	F43 Barking				

CUB 1-5
Locomotors (Command Unit Body)   1987

1	Reserve	3	F20 Stratford EC	5	E20 Lewisham EC
2	G20 Wembley WC	4	E20 Lewisham EC		

DCB 2-7
Locomotors (Damage Control Unit Body)   1984

2	H38 Sutton	5	Reserve	7	A34 Edmonton
4	F43 Barking				

EUB3
Torton Bodies (Exhibition Unit Body)   1991   Fire Prevention

HLB 1-12   Wadham Stringer (Hose Laying Body)   1982-83

1	G40 Hayes	6	A34 Edmonton	9	F43 Barking	
2	G40 Hayes	7	E43 Beckenham	11	H38 Sutton	
3	H38 Sutton	8	F43 Barking	12	E43 Beckenham	
5	A34 Edmonton					

ICB1   Locomotors/Penman (Incident Conference Unit)   1989   A34 Edmonton

---

MIB 1-5   Saxon Sanbec Sanbec (BA Major Incident Unit Body)   1994

1	Brigade Reserve	3	E43 Beckenham	5	H38 Sutton
2	A34 Edmonton	4	G40 Hayes		

OSB 1-4   Saxon Sanbec (Operational Support Unit)   1995

1	E43 Beckenham	3	F43 Barking	4	Reserve
2	G40 Hayes				

DEMOUNTABLE BODIES FOR MEDIUM CHASSIS UNITS (MDC's)

DSB1   Ray Smith/Ruislip Workshops (Dropside Body)   1986/93   Ruislip Works (Ex FCB9)

FIB 1-5   Ray Smith/Wadham Stringer (Fire Investigation Unit Body)   1984

1	A27 Clerkenwell	3	F44 East Ham	5	H21 Clapham
2	E38 New Cross	4	G26 Acton		

PCB 1-16   Ray Smith/Wadham Stringer (Personnel Carrier Body)   1982-83

1	West Training	6	F20 Stratford EC	12	E20 Lewisham SC
2	E20 Lewisham SC	8	West Training	14	Training Centre
3	F20 Stratford EC	9	F20 Stratford EC	15	Training Centre
4	E20 Lewisham SC	10	E20 Lewisham SC	16	Training Centre
5	E20 Lewisham SC	11	West Training		

RDB1   Ray Smith/Wadham Stringer (Research & Development Body)   1983   Personnel Section

---

VB 1-9   Ray Smith/CFE (Van Body)   1981

2	F20 Stratford EC	6	G20 Wembley WC	8	Protective Equipment Group
3	E20 Lewisham SC	7	E20 Lewisham SC	9	Protective Equipment Group
5	G20 Wembley WC				

VB 10-29   Ray Smith/Wadham Stringer (Van Body)   1982-83

10	E20 Lewisham SC	16	G20 Wembley WC	23	F20 Stratford EC
11	G20 Wembley WC	17	G20 Wembley WC	24	E20 Lewisham SC
12	F20 Stratford EC	18	F20 Stratford EC	26	G20 Wembley WC
13	E20 Lewisham SC	19	F20 Stratford EC	27	Headquarters
14	Headquarters	21	E20 Lewisham SC	28	F20 Stratford EC
15	G20 Wembley WC	22	F20 Stratford EC	29	Training Centre

VB30	Ray Smith/Hawson Garner (Van Body)	1988	Ruislip Works
VB31	Ray Smith/Hawson Garner (Van Body)	1988	Fleet Management Dept.
VB32	Hawson (Van Body)	1994	Training Centre
VB33	Garner (Van Body)	1994	E20 Lewisham SC

LOTHIAN & BORDERS FIRE BRIGADE

Lothian and Borders Fire Brigade, Lauriston Place, Edinburgh EH3 9DE

WSG106W	Dodge G1313	Carmichael	WrL	1980	Reserve
WSG107W	Dodge G1313	Carmichael	WrL	1980	Reserve
WSG109W	Dodge G1313	Carmichael	WrL	1980	Reserve
WSG110W	Dodge G1313	Carmichael	WrL	1980	Training School
WSG112W	Dodge G1313	Carmichael	WrL	1980	Training School
HSG735X	Dodge G1313	Fulton & Wylie	WrL	1980	Reserve
HSG737X	Dodge G1313	Fulton & Wylie	WrL	1980	Reserve
HSG738X	Dodge G1313	Fulton & Wylie	WrT	1980	39 Peebles
HSG740X	Dodge G1313	Fulton & Wylie	WrL	1980	Reserve
HSG741X	Dodge G1313	Penman	FST	1980	52 Marionville
MFS104X	Dennis F125	Dennis/Simon Snorkel SS263	HP	1982	Reserve
NSX341Y	Dodge G13C	Mountain Range	WrL	1983	Training Centre
NSX347Y	Dodge G13C	Mountain Range	WrT	1983	Reserve
NSX348Y	Dodge G13C	Mountain Range	WrT	1983	Reserve
A318ASF	Shelvoke & Drewry WY	Angloco/Metz DL30	TL	1984	51 Crewe Toll
A49EMS	Dodge G13C	Mountain Range	WrL	1984	35 Hawick
A50EMS	Dodge G13C	Mountain Range	WrT	1984	35 Hawick
A51EMS	Dodge G13C	Mountain Range	WrL	1984	34 Galashiels
A52EMS	Dodge G13C	Mountain Range	WrL	1984	Reserve
A451LND	Dodge S66	Mountain Range	ET	1984	54 Bathgate
A494WDM	Bedford TKG	Saxon Sanbec SVB	CU	1984	32 Liberton
B631JFS	Dodge G13C	Mountain Range	WrL	1985	Reserve
B632JFS	Dodge G13C	Mountain Range	WrT	1985	Reserve
B633JFS	Dodge G13C	Mountain Range	WrL	1985	46 Kelso
B634JFS	Dodge G13C	Mountain Range	WrL	1985	Training School
C226RSC	Dodge G13C	Alexander	WrL	1986	Reserve
C227RSC	Dodge G13C	Alexander	WrT	1986	54 Bathgate
C228RSC	Dodge G13C	Alexander	WrT	1986	65 Haddington
C229RSC	Dodge G13C	Alexander	WrT	1986	68 North Berwick
C793USX	Dodge G13C	Mountain Range	WrL/R	1984	62 Linlithgow
C794USX	Dodge G13C	Mountain Range	WrL/R	1984	49 Eyemouth
C795USX	Dodge G13C	Mountain Range	WrL/R	1984	38 WestLinton
D467BSC	Dodge G16C	Saxon/Simon Snorkel SS263	HP	1987	50 McDonald Road
D493BSC	Dodge G13C	Mountain Range	WrL/R	1984	69 Dunbar
D494BSC	Dodge G13C	Mountain Range	WrL/R	1984	65 Haddington
D495BSC	Dodge G13C	Mountain Range	WrL/R	1984	39 Peebles
D927CRF	Bedford TL	Excalibur CBK	ET	1987	34 Galashiels
E106MSC	Dodge G13C	Mountain Range	WrL/R	1984	46 Kelso
E107MSC	Dodge G13C	Mountain Range	WrL/R	1984	45 Jedburgh
E108MSC	Dodge G13C	Mountain Range	WrL/R	1984	47 Coldstream
F903USX	Renault-Dodge G13C	Excalibur CBK	WrL/R	1989	42 Newcastleton
F904USX	Renault-Dodge G13C	Excalibur CBK	WrL/R	1989	37 Penicuik
F905USX	Renault-Dodge G13C	Excalibur CBK	WrL/R	1989	58 South Queensferry
F906USX	Renault-Dodge G13C	Excalibur CBK	WrL/R	1989	Reserve
F907USX	Renault-Dodge G13C	Excalibur CBK	WrL/R	1989	68 North Berwick
G39DSF	Renault-Dodge G16C	Carmichael/Magirus DL30	TL	1990	30 Tollcross
G871FFS	Renault-Dodge G13C	Fulton & Wylie	WrL	1990	59 Broxburn
G238FSC	Renault-Dodge G13C	Fulton & Wylie	WrL	1990	48 Duns
G239FSC	Renault-Dodge G13C	Fulton & Wylie	WrL/R	1990	64 Tranent
G240FSC	Renault-Dodge G13C	Fulton & Wylie	WrL	1990	40 Innerleithen
G241FSC	Renault-Dodge G13C	Fulton & Wylie	WrL	1990	67 East Linton
H94NSX	Volvo FL6.17	Mountain Range	WrL/R	1991	61 Whitburn
H531PSX	Scania G93M-250	Mountain Range	WrL/R	1991	Training School
J319USC	Mercedes-Benz 1222F	Mountain Range	WrL/R	1991	60 West Calder
J851WSC	Renault-Dodge S46	Renault/L&BFB	CU/GPV	1992	32 Liberton

Opposite, top: **One of the last aerial appliances to be built with a full-size crew cab is G39DSF of Lothian & Borders Fire Brigade. The rare type is shown in this picture of the appliance which is based at Tollcross.** *Andrew Fenton*

Opposite, bottom: **Page 92 shows the picture of the Volvo used in comparative trails with Lothian and Borders. Shown here is one of the Scania P93Ms which are now being supplied as the standard pump. M137XSF has coachwork by Emergency One.** *Andrew Fenton*

The Fire Brigade Handbook

A Volvo FL6 was procured in 1991 for evaluation against Scania and Mercedes-Benz Water Tender Ladders. The Volvo and Mercedes machines remain as the solitary examples of each in the fleet, while Scanias products have been adopted for the brigades needs. H94NSX was photographed at Marionville fire station in Edinburgh shortly before transfer to the retained Whitburn Station.
Alistair MacDonald

K959DSC	Scania G93M-250	Emergency One	WrL/R	1992	55 Livingston
K960DSC	Scania G93M-250	Emergency One	WrL/R	1992	41 Selkirk
K961DSC	Scania G93M-250	Emergency One	WrL/R	1992	53 Musselburgh
K962DSC	Scania G93M-250	Emergency One	WrL/R	1992	42 Melrose
K963DSC	Scania G93M-250	Emergency One	WrL/R	1992	35 Hawick
K964DSC	Scania G93M-250	Emergency One	WrL/R	1992	33 Dalkeith
L492JSC	Scania G93M-250	Emergency One	WrL/R	1992	54 Bathgate
L281NSC	Scania G93M-250	Emergency One	WrL/R	1994	36 Newcraighall
L283NSC	Scania G93M-250	Emergency One	WrL/R	1994	50 McDonald Road
L284NSC	Scania G93M-250	Emergency One	WrL/R	1994	30 Tollcross
M472SSX	Scania G93M-250	Emergency One	HRU	1994	36 Newcraighall
M473SSX	Scania P113M-280	Bedwas/Simon Snorkel ST290S	ALP	1995	31 Sighthill
M133XSF	Scania G93M-250	Emergency One	WrL/R	1995	52 Marionville
M134XSF	Scania G93M-250	Emergency One	WrL/R	1995	34 Galashiels
M135XSF	Scania G93M-250	Emergency One	WrL/R	1995	32 Liberton
M136XSF	Scania G93M-250	Emergency One	WrL/R	1995	30 Tollcross
M137XSF	Scania G93M-250	Emergency One	WrL/R	1995	50 McDonald Road, Edinburgh
N303FSG	Scania G93M-250	Emergency One	WrL/R	1996	31 Sighthill
N304FSG	Scania G93M-250	Emergency One	WrL/R	1996	51 Crewe Toll
N305FSG	Scania G93M-250	Emergency One	WrL/R	1996	31 Sighthill
N306FSG	Scania G93M-250	Emergency One	WrL/R	1996	51 Crewe Toll
P867LSC	Scania G93M-250	Penman	ISU	1996	

Note: LFS400T was converted from a WrL.

MERSEYSIDE FIRE BRIGADE

Merseyside Fire Brigade, Hatton Gardens, Liverpool L3 2AD

	OYH414R	Bedford TKG	HCB-Angus CSV/MFB(1990)	WrL	1977	Driving School
1193	TTJ317V	Dodge S56	Cocker	CaV	1980	W1 Birkenhead
1198	XKB927W	Ford D0907	Merseyside FB/HIAB	MPU	1980	Reserve
1203	ACM407X	Dennis RS133	Dennis	WrL	1982	Driving School
1205	ACM409X	Dennis RS133	Dennis	WrL	1982	Training Centre
1207	ACM413X	Dennis F125	Dennis/Simon Snorkel SS263	HP	1982	Reserve/Driving School
1211	DWM607Y	Ford Cargo 16.250	Malmo/Merseyside FB	FoT	1983	W1 Birkenhead
1216	DWM612Y	Dennis RS133	Dennis	WrL	1983	Training Centre
1219	A50HKC	Ford Cargo 16.250	Malmo/Merseyside FB	FoT	1984	Training Centre
1221	A48HKC	Dennis F125	Dennis/Simon Snorkel SS65	HP	1984	Reserve
1226	B920KWM	Dennis RS133	Dennis	WrL	1984	Training Centre
1229	B923KWM	Dennis F125	Dennis/Simon Snorkel SS65	HP	1985	Reserve
1232	GCA684X	Land Rover Series III 109	Merseyside FB(1986)	L4P	1982	W5 West Kirby
1233	D174SKC	Dennis SS135	Carmichael	WrL	1987	Driving School
1234	D175SKC	Dennis SS135	Carmichael	WrL	1987	Reserve
1235	E494UKF	Dennis SS135	Carmichael	WrL	1987	Reserve
1236	D177SKC	Dennis SS135	Carmichael	WrL	1987	Reserve
1237	D178SKC	Renault-Dodge G13C	Excalibur CBK	WrL	1987	Driving School
1238	D179SKC	Renault-Dodge G13C	Excalibur CBK	WrL	1987	Driving School
1241	B593NPL	Dennis F125	John Dennis/Camiva EPA30	TL	1985	C1 Kirkdale
1242	GCA683X	Land Rover Series III 109	Merseyside F.B.(1986)	L4V	1982	N7 Southport
1243	E496UKF	Renault-Dodge G13C	Carmichael	CU	1987	Supplies
1244	E497UKF	Dennis SS135	Mountain Range	WrL	1988	Training Centre
1245	E498UKF	Dennis SS135	Mountain Range	WrL	1988	Reserve
1246	F922YCM	Volvo FL6.17	Angloco/Metz DL30	TL	1989	Reserve
1247	F930YCM	Dennis SS135	Mountain Range	WrL	1989	Reserve
1248	F931YCM	Dennis SS135	Mountain Range	WrL	1989	Reserve
1249	F932YCM	Dennis SS135	Mountain Range	WrL	1989	Reserve
1250	F933YCM	Dennis SS135	Mountain Range	WrL	1989	Reserve
1251	F934YCM	Dennis SS135	Mountain Range	WrL	1989	Reserve
1252	F935YCM	Dennis SS135	Mountain Range	WrL	1989	Reserve
1253	G394EKA	Dennis SS135	Mountain Range	WrL	1990	W1 Birkenhead
1254	G395EKA	Dennis SS135	Mountain Range	WrL	1990	W3 Heswall
1255	G396EKA	Dennis SS135	Mountain Range	WrL	1990	W2 Bebington
1256	G397EKA	Dennis SS135	Mountain Range	WrL	1989	N4 Crosby
1257	G398EKA	Dennis SS135	Mountain Range	WrL	1990	E1 St.Helens
1258	G399EKA	Dennis SS135	Mountain Range	WrL	1990	E6 Kirkby
1259	G400EKA	Dennis SS135	Mountain Range	WrL	1990	C4 Low Hill
1260	G401EKA	Dennis SS135	Mountain Range	WrL	1990	C3 City Centre, Liverpool
1261	H149GBG	Dennis SS237	Excalibur CBK	WrL	1991	W2 Bebington
1262	H150GBG	Dennis SS237	Excalibur CBK	WrL	1991	W6 Wallasey

In 1992, Merseyside Fire Brigade purchased a Dennis Rapier for evaluation and this appliance spent its time at Derby Lane, Liverpool, now renamed Old Swan. Three further chassis types are due in 1997 with products from Volvo, Dennis and Mercedes-Benz, again for evaluation.
Tim Ansell

1263	H151GBG	Dennis SS237	Excalibur CBK	WrL	1991	S5 Garston
1264	H152GBG	Dennis SS237	Excalibur CBK	WrL	1991	N2 Aintree
1265	H153GBG	Dennis SS237	Excalibur CBK	WrL	1991	E3 Huyton
1266	H154GBG	Dennis SS237	Excalibur CBK	WrL	1991	W4 Upton
1267	H155GBG	Dennis SS237	Excalibur CBK	WrL	1991	S8 Old Swan
1268	H156GBG	Dennis SS237	Excalibur CBK	WrL	1991	C4 Low Hill
1269	H933JKD	Dennis SS237	Excalibur CBK	WrL	1991	C1 Kirkdale
1270	H934JKD	Dennis SS237	Excalibur CBK	WrL	1991	N7 Southport
1271	H935JKD	Dennis SS237	Excalibur CBK	WrL	1991	S9 Belle Vale
1272	H936JKD	Dennis SS237	Excalibur CBK	WrL	1991	W1 Birkenhead
1273	H937JKD	Dennis SS237	Excalibur CBK	WrL	1991	N1 Bootle & Netherton
1274	H938JKD	Dennis SS237	Excalibur CBK	WrL	1991	W5 West Kirby
1275	H939JKD	Dennis SS237	Excalibur CBK	WrL	1991	S1 Allerton
1276	J252KWM	Dennis Rapier TF202	Excalibur CBK	WrL	1992	S8 Old Swan
1277	H157GBG	Volvo FL10	Angloco/Bronto Skylift 28-2Ti	CPL	1991	N7 Southport
1278	H158GBG	Volvo FL10	Angloco/Bronto Skylift 28-2Ti	CPL	1991	C3 City Centre, Liverpool
1279	H653JKF	Dennis DFS237	Penman Multilift	PM	1991	Reserve
1280	H654JKF	Dennis DFS237	Penman Multilift	PM	1991	Reserve
1281	J256KWM	Dennis DFS237	Penman Multilift	PM	1992	C1 Kirkdale
1282	J257KWM	Dennis DFS237	Penman Multilift	PM	1992	W1 Birkenhead
1283	K851NEM	Dennis SS237	Excalibur CBK	WrL	1993	E1 St.Helens
1284	K852NEM	Dennis SS237	Excalibur CBK	WrL	1993	E6 Kirkby
1285	K853NEM	Dennis SS237	Excalibur CBK	WrL	1993	C1 Kirkdale
1286	K854NEM	Dennis SS237	Excalibur CBK	WrL	1993	E2 Newton Le Willows
1287	K855NEM	Dennis SS237	Excalibur CBK	WrL	1993	C3 City Centre, Liverpool
1288	K856NEM	Dennis SS237	Excalibur CBK	WrL	1993	W2 Bebington
1289	K857NEM	Dennis SS237	Excalibur CBK	WrL	1993	N3 Croxteth
1290	K858NEM	Dennis SS237	Excalibur CBK	WrL	1993	N1 Bootle & Netherton
1291	K945OHF	Dennis DFS237	Penman Multilift	PM	1993	C1 Kirkdale
1292	K946OHF	Dennis DFS237	Penman Multilift	PM	1993	E1 St.Helens
1293	K474OKB	Volvo FL10	Angloco/Bronto Skylift 28-2Ti	CPL	1993	E1 St.Helens
1294	K475OKB	Volvo FL10	Angloco/Bronto Skylift 28-2Ti	CPL	1993	W1 Birkenhead
1295	L38SFY	Dennis DFS241	Cargotec Multilift	PM	1994	W1 Birkenhead
1296	L39SFY	Dennis DFS241	Cargotec Multilift	PM	1994	Command & Control
1297	L243STJ	Dennis SS243	Excalibur CBK	WrL	1994	N4 Crosby
1298	L243STJ	Dennis SS243	Excalibur CBK	WrL	1994	N7 Southport
1299	L243STJ	Dennis SS243	Excalibur CBK	WrL	1994	S6 Toxteth
1300	L243STJ	Dennis SS243	Excalibur CBK	WrL	1994	S6 Toxteth
1301	L243STJ	Dennis SS243	Excalibur CBK	WrL	1994	S2 Speke
1302	M771WKC	Dennis SS243	Excalibur CBK	WrL	1994	E5 Eccleston
1303	M772WKC	Dennis SS243	Excalibur CBK	WrL	1994	W6 Wallasey
1304	M773WKC	Dennis SS243	Excalibur CBK	WrL	1994	N7 Southport
1305	M774WKC	Dennis SS243	Excalibur CBK	WrL	1994	N6 Formby
1306	M775WKC	Dennis SS243	Excalibur CBK	WrL	1994	N3 Croxteth
1307	M776WKC	Dennis SS243	Excalibur CBK	WrL	1994	E4 Whiston
1308	M232YBG	Volvo FS7-18	Cargotec Multilift	PM	1995	E1 St Helens
1309	M233YBG	Volvo FS7-18	Cargotec Multilift	PM	1995	W6 Wallasey
1310	N65DEM	Volvo FL10	Angloco/Bronto Skylift F32-HDI	CPL	1996	N3 Croxteth
5001	Pod	Cocker	Special Rescue Unit	SRU	1992	E1 St.Helens
5002	Pod	Bence	Special Rescue Unit	SRU	1992	C1 Kirkdale
5003	Pod	Leicester Carriage	Command & Control Unit	CCU	1991	Command & Control Centre, Liverpool
5004	Pod	Bence	B.A.Support Unit	BASU	1992	W1 Birkenhead
5005	Pod	Bence	Operational Support Unit	OSU	1992	W1 Birkenhead
5006	Pod	Bence	Operational Support Unit	OSU	1992	W1 Birkenhead
5007	Pod	Bence	Damage Control Unit	DCU	1992	E1 St Helens
5008	Pod	Leicester Carriage	Chemical Incident Unit	CIU	1992	C1 Kirkdale
5009	Pod	Bence	Marine Unit	MarU	1992	C1 Kirkdale
5010	Pod	Devcoplan	Hose Laying Lorry	HL	1992	C1 Kirkdale
5011	Pod	Devcoplan	Hose Laying Lorry	HL	1992	E1 St.Helens
5012	Pod	Devcoplan	Hose Laying Lorry	HL	1993	W1 Birkenhead
5013	Pod	Bence	B.A.Training Pod	BA(T)	1994	C1 Kirkdale
5014	Pod	Bence	B.A.Training Pod	BA(T)	1994	E3 Huyton
5015	Pod	Bence	B.A.Training Pod	BA(T)	1994	S9 Belle Vale
5016	Pod	Bence	Exhibition Unit	ExU	1994	W6 Wallasey
5017	Pod	Bence	Bulk Foam Unit	BFU	1995	W1 Birkenhead
5018	Pod	Bence	Bulk Foam Unit	BFU	1995	E1 St Helens
5019	Pod	Bence	Bulk Foam Unit	BFU	1995	C1 Kirkdale
5020	Pod	Bence	Bulk Foam Unit	BFU	1995	W6 Wallasey
5021	Trailer	Concept BB	Display Trailer	ExU	1995	W3 Heswall
5022	Trailer	Concept BB	Display Trailer	ExU	1995	E5 Eccleston

Note: OYH414R was acquired from The Fire Service College, Moreton-In-Marsh.

Merseyside's Volvo appliances are currently aerial appliances or Prime Movers. St Helens fire station have one of each and seen here are Prime Mover 1308 (M232YBG) with Special Rescue pod 5001, beside is 1293 (K474OKB) a Combined Platform Ladder appliance (CPL). *Andy Daley*

Merseyside Fire Brigade now have five Volvo FL10s with Angloco bodywork and Bronto Skylift booms. The newest appliance has the latest F32HDT booms, the other four having the 28-2Ti model. Numerically the first of the type, 1277 (H157GBG) is based at Croxteth. *Andy Daley*

MID & WEST WALES FIRE BRIGADE

Mid & West Wales Fire Brigade, Ucheldir, College Road, Carmarthen, SA31 3EF

(C - acquired from Powys FB; D - acquired from Dyfed and W - acquired from West Glamorgan)

P	AFO144B	Land Rover Series II 88	Brecon & Radnor Joint FB	L4T	1964	N15 Llanwrtyd Wells
P	DFO855G	Land Rover Series II 88	Brecon & Radnor Joint FB	L4T	1969	N17 Presteigne
D	EGP136J	Daimler Fleetline CRG6LXB	Park Royal	DisU	1971	Headquarters
P	SVJ241S	Land Rover Series III 109	Saxon Sanbec	L4P	1977	N7 Llanidloes
P	TCJ557S	Ford A0610	Cheshire Fire Engineering	LRT	1978	N10 Brecon
P	TCJ558S	Bedford TK1260	Carmichael	WRT	1978	N Reserve
P	TCJ559S	Bedford TK1260	Carmichael	WRT	1978	N Reserve
P	YKM86S	Land Rover Series III 109	Spenklin Branbridge	L4P	1978	N9 Rhayader
D	VBX122T	Dennis D	Dennis	WrL	1978	Caldy Island Volunteers
P	BFO308V	Ford A0610	Angloco	LRT	1979	N1 Newtown
P	XUD565V	Bedford TK1260	HCB-Angus	WrL	1979	Reserve
P	CCJ270V	Bedford TK1260	Carmichael	WrL	1980	N7 Llanidloes
D	DDE871W	Dennis D	Dennis	WrL	1980	N Reserve
D	Pod	Carmichael Benson	B.A. Training Pod	BA(T)	1981	W5 Ammanford
P	TJO728X	Bedford TK1260	HCB-Angus	WrL	1981	N6 Llanfyllin
D	JDE805X	Dodge G13	Cheshire Fire Engineering	WrL	1982	W Training School
D	LBX805X	Dodge G7575	Cheshire Fire Engineering	FEV	1982	W12 Milford Haven
P	MCY402X	Bedford TK1260	Cheshire Fire Engineering	WrL	1982	N2 Machynlleth
P	NTU515Y	Bedford TK1260	Mountain Range	WrL	1983	N16 Hay on Wye
P	OMA125Y	Bedford TK1260	Saxon Sanbec	WrL	1983	N Reserve
P	OMA126Y	Bedford TK1260	Saxon Sanbec	WrL	1983	N4 Welshpool
P	ONP785Y	Dennis RS133	Dennis	WrL	1983	N Training School
W	SEP544Y	Dennis SS133	Dennis	WrL	1983	S Reserve
D	Trailer	Lynton Trailers	Control Unit	CU	1984	W16 Pembroke Dock
D	Pod	Carmichael/Hoskins(1989)	Foam Tender	FoT	1984	W16 Pembroke Dock
W	PCY814Y	Dodge S66C	Angloco	RT	1983	S Reserve
W	A721WTH	Dennis DS151	Dennis	WrL	1984	S Reserve
P	A621VTU	Bedford TK1260	Saxon Sanbec	WrL	1984	N5 Llanfair Caereinion
P	A622VTU	Bedford TK1260	Saxon Sanbec	WrL	1984	N15 Llanwrtyd Wells
P	A623VTU	Bedford TK1260	Saxon Sanbec	WrL	1984	N10 Brecon
P	B645VUD	Bedford TK1260	HCB-Angus	WrL	1984	N Training
P	B899DMA	Bedford TK1260	Saxon Sanbec	WrL	1985	N12 Crickhowell
W	B339CCY	Dennis DS151	Dennis	WrL	1985	S Reserve
P	B478DMB	Land Rover 110	Saxon Sanbec	L4P	1985	N6 Llanfyllin
P	B479DMB	Land Rover 110	Saxon Sanbec	L4P	1985	N8 Knighton
P	B623DTU	Land Rover 110	Saxon Sanbec	L4P	1985	N11 Abercrave
D	B561WDE	Dodge G13	Mountain Range	WrL	1985	W Reserve
D	B556XBX	Dennis SS135	Dennis	WrL	1985	W Reserve
D	B298YDE	Dennis SS135	Dennis	WrL	1985	W Reserve
D	B299YDE	Dennis SS135	Dennis	WrL	1985	W Reserve
P	C447HMA	Bedford TK1260	Saxon Sanbec	WrL	1985	N9 Rhayader
P	C688VVT	Bedford M1120 4x4	Saxon Sanbec	WRC	1986	N14 Builth Wells
D	C205XVJ	Land Rover 110	Excalibur CBK	L4P	1986	N12 Crickhowell
W	C233JWN	Dodge G13C	Angloco	WrL	1986	S7 Reynoldston
W	C234JWN	Dodge G13C	Angloco	WrL	1986	S4 Port Talbot
W	C641JTH	Dodge G13C	Angloco	WrL	1986	S3 Cymmer
D	D683FDE	Renault-Dodge G16	Saxon/Simon Snorkel SS263	HP	1986	N21 Aberystwyth
D	D303GBX	Land Rover 110	Land Rover/Saxon Sanbec	L4P	1986	N19 Borth
P	D751LUH	Bedford TL1200	Excalibur CBK	WrL	1986	N17 Presteigne
D	D693GBX	Dennis SS135	Angloco	WrL	1987	N21 Aberystwyth
D	D694GBX	Dennis SS135	Angloco	WrL	1987	B21 Aberystwyth
D	D404HDE	Dennis DS151	Angloco	PL	1987	W18 Crymych
D	D405HDE	Dennis DS151	Angloco	PL	1987	N22 Tregaron
D	D30JBX	Ford Transit	Ford	PCV	1987	FP Support
D	E263LBX	Ford Transit Artic	Lynton Trailers	TU	1987	W16 Pembroke Dock
D	E750MDE	Scania G92M	Angloco/Metz DL30	TL	1987	W11 Haverfordwest
W	E962RKG	Scania G92M	Carmichael	WrL	1987	S Training
W	E963RKG	Scania G92M	Carmichael	WrL	1987	S10 Seven Sisters
W	E681SBO	Scania 92M	Carmichael	WrL	1987	S2 Glynneath
P	E541ATU	Bedford M1120 4x4	Excalibur CBK	WRC	1988	N1 Newtown
W	E793AWO	Scania 93M-210	Lacre PDE Rolonof III	PM	1988	S8 Morriston, Swansea
W	F506YTX	Scania 92M	Carmichael	WrL	1988	S12 Pontardulais

Mid & West Wales Fire Brigade was formed on 1st April 1996 by amalgamating the former Dyfed, Powys and West Glamorgan Brigades together. The former Powys brigade, being particularly rural, had a requirement for just one aerial appliance. Between 1971 and 1994 it was a Bedford Hydraulic Platform, but it was replaced by this Volvo/Bronto Skylift M163PFO. *Robert Smith*

The last new appliance for Dyfed County Fire Brigade were Dennis Sabres bodied by Carmichael International. The first pair were allocated to Pembroke Dock Fire Station where M630CDE was photographed when new. *Paul Measday*

W	F507YTX	Scania 93M	Carmichael	WrL	1988	S11 Gorseinon
D	Pod	Hoskins	Canteen Van Pod	CaV	1988	W16 Pembroke Dock
D	Pod	Hoskins	Hazardous Substance Pod	HS	1988	W1 Llanelli
W	F367YFX	Ford Transit/County 4x4	HCB-Angus	L4P	1988	S6 Sketty Green, Swansea
W	F368YFX	Ford Transit/County 4x4	HCB-Angus	L4P	1988	S9 Pontardawe
D	F417SBX	Volvo FL6.14	Angloco	PL	1989	W Training Centre
D	F425TDE	Volvo FL6.14	Angloco	WrL	1989	W11 Haverfordwest
D	F426TDE	Dennis SS135	Saxon Sanbec	WrL	1989	W12 Milford Haven
D	F427TDE	Dennis SS135	Saxon Sanbec	WrL	1989	W12 Milford Haven
D	F428TDE	Scania G93ML-250	Angloco	FoT	1989	W12 Milford Haven
D	F429TDE	Scania G93ML	Hoskins	PM	1989	W16 Pembroke Dock
W	G958BRU	Mercedes-Benz 917AF 4x4	HCB-Angus	RT	1989	S8 Morriston, Swansea
W	G607GHB	Scania 93H	Fulton & Wylie	WrL	1989	S4 Port Talbot
W	G608GHB	Scania 93H	Fulton & Wylie	WrL	1989	S9 Pontardawe
D	G987XBX	Land Rover 110	Dyfed CFB	L4R	1989	W2 Carmarthen
D	G896YDE	Volvo FL6.14	Angloco	WrL	1989	W11 Haverfordwest
P	G839KAW	Mercedes-Benz 1120F	Excalibur CBK	WrL	1990	N11 Abercrave
P	G840KAW	Mercedes-Benz 1120F	Excalibur CBK	WrL	1990	N1 Newtown
D	G646ABX	Scania G93M	Angloco/Metz DLK30	TL	1990	W2 Carmarthen
D	G647ABX	Dennis SS237	Saxon Sanbec	WrL	1990	W2 Carmarthen
D	G648ABX	Dennis SS237	Saxon Sanbec	WrL	1990	W2 Carmarthen
D	G649ABX	Dennis SS237	Saxon Sanbec	WrL	1990	W5 Ammanford
W	G429KBO	Scania G93M-250	Angloco/Metz DLK30	TL	1990	S8 Morriston, Swansea
W	G216YTW	Scania G93M-250	Bedwas/Simon Snorkel SS220	HP	1990	S8 Morriston, Swansea
W	H549MDW	Scania G93M-210	Lacre PDE Rolonof III	PM	1990	S8 Morriston, Swansea
D	H517FBX	Iveco Daily 49.10	Days	TU	1990	W1 Llanelli
D	Trailer	Lynton Trailers	Information Unit	IU	1990	W1 Llanelli
P	H162PNT	Mercedes-Benz 1120F	Excalibur CBK	WrL	1990	N8 Knighton
P	H163PNT	Mercedes-Benz 1120F	Excalibur CBK	WrL	1990	N13 Llandrindod Wells
D	H616EEJ	Mercedes-Benz 1120AF	Carmichael/HIAB	ET	1990	N2 Aberystwyth
D	H623EBX	Mercedes-Benz 1120AF	Carmichael/HIAB	ET	1990	W1 Llanelli
D	H949EDE	Mercedes-Benz 1120AF	Carmichael/HIAB	ET	1990	W11 Haverfordwest
D	H950EDE	Mercedes-Benz 609D	Hoskins	MRU	1991	W12 Milford Haven
D	Boat	Avon Searider 5.5M	Boat and trailer	IRBt	1991	W12 Milford Haven
D	H684FBX	Dennis SS239	Carmichael	WrL	1991	W8 Llandysul
D	H685FBX	Dennis SS239	Carmichael	WrL	1991	W10 Llandovery
D	H686FBX	Dennis SS239	Carmichael	WrL	1991	W6 Tumble
D	H687FBX	Dennis SS239	Carmichael	WrL	1991	W7 Newcastle Emlyn
D	H697FEJ	Dennis DS151	Excalibur CBK	PL	1991	N24 Aberaeron
D	H577GDE	Dennis DS151	Excalibur CBK	PL	1991	W14 St.Davids
D	H578GDE	Dennis SS239	Carmichael	WrL	1991	W17 Tenby
D	H579GDE	Dennis SS239	Carmichael	WrL	1991	W17 Tenby
W	H562MDW	Scania P93H	HCB-Angus	WrL	1991	S8 Morriston, Swansea
W	H496STH	Ford Transit/County 4x4	Carmichael	L4P	1991	S4 Port Talbot
W	H497STH	Ford Transit/County 4x4	Carmichael	L4P	1991	S8 Morriston, Swansea
D	J930MBX	Scania P93M-250	Boniface/Interstater II	HRU	1991	W2 Carmarthen
W	J654STX	Scania G92M-250	Excalibur CBK	WrL	1992	S6 Sketty Green, Swansea
P	J544WUX	Mercedes-Benz 1120F	Excalibur CBK	WrL	1992	N1 Newtown
P	J545WUX	Mercedes-Benz 1120F	Excalibur CBK	WrL	1992	N10 Brecon
D	J829KEJ	Dennis SS239	Excalibur CBK	WrL	1992	N20 Lampeter
D	J830KEJ	Dennis SS239	Excalibur CBK	WrL	1992	N20 Lampeter
D	J940NBX	Dennis SS239	Excalibur CBK	WrL	1992	W4 Pontyates
D	J941NBX	Dennis SS239	Excalibur CBK	WrL	1992	W1 Llanelli
D	J942NBX	Dennis SS239	Excalibur CBK	WrL	1992	W1 Llanelli
D	Pod	Leicester Carriage	B A Training Unit	BA(T)	1992	W1 Llanelli
W	K428YDW	Scania G93M-250	Lacre PDE Rolonof III	PM	1992	S8 Morriston, Swansea
P	K945BUJ	Mercedes-Benz 1120F	Excalibur CBK	WrL	1992	N4 Welshpool
P	K946BUJ	Mercedes-Benz 1120F	Excalibur CBK	WrL	1992	N3 Montgomery
W	K516BTX	Scania G93M-250	HCB-Angus	WrL	1993	S8 Morriston, Swansea
W	K517BTX	Scania G93M-250	HCB-Angus	WrL	1993	S5 Parc Tawe, Swansea
D	K141SBX	Volvo FL7.18	Powell Duffryn Rolonof	PM	1993	W1 Llanelli
D	K749SBX	Dennis DS155	Excalibur CBK	PL	1993	W9 Llandeilo
D	K852TDE	Dennis DS155	Carmichael	PL	1993	W20 Whitland
D	K853TDE	Dennis DS155	Excalibur CBK	PL	1993	W13 Fishguard
D	Pod	Leicester Carriage/Dairy Products	Water Carrier	WrC	1993	W1 Llanelli
W	Pod	HCB-Angus	Foam Tender	FoT	1993	S8 Morriston
P	L925GUX	Mercedes-Benz 1124F	Excalibur CBK	WrL	1993	N2 Machynlleth
P	L926GUX	Mercedes-Benz 1124F	Excalibur CBK	WrL	1993	N18 Talgarth
D	L431SEJ	Dennis DS155	Carmichael International	PL	1994	N23 New Quay
D	L149XBX	Dennis DS155	Carmichael International	PL	1994	W3 Kidwelly
D	L493YDE	Dennis DS155	Carmichael International	PL	1994	W19 Cardigan
D	L494YDE	Dennis DS155	Carmichael International	PL	1994	W15 Narbeth
P	M163PFO	Volvo FL10	Angloco/Bronto Skylift 24-HDT	HP	1994	N13 Llandrindod Wells

The three constituent brigades that formed Mid & West Wales Fire Brigade had different preferences for their new pumping appliances. Dyfed bought Dennis, Powys had settled on Mercedes-Benz, and West Glamorgan showed allegiance to Scania products. Shown here is former West Glamorgan's M595KUH, a P93M bodied by Saxon Sanbec. *Paul Measday*

P	M353PFO	Mercedes-Benz 1124F	Excalibur GM2700	WrL	1994	N13 Llandrindod Wells
P	M354PFO	Mercedes-Benz 1124F	Excalibur GM2700	WrL	1994	N14 Builth Wells
D	M629CDE	Dennis Sabre TSD233	Carmichael International	WrL	1995	W16 Pembroke Dock
D	M630CDE	Dennis Sabre TSD233	Carmichael Interational	WrL	1995	W16 Pembroke Dock
W	M595KUH	Scania P93M-250	Saxon Sanbec	WrL	1995	S1 Neath
	N423PDW	Scania P93M-250	Saxon Sanbec	WrL	1996	
	N316AEJ	Dennis Sabre S411	Carmichael Interational	WrL	1996	N2 Aberystwyth
	N418JBX	Dennis Sabre S411	Carmichael Interational	WrL	1996	W5 Ammanford
	N	Mercedes-Benz 1124F	Excalibur CBK	WrL	1996	N1 Newtown
	N	Mercedes-Benz 1124F	Excalibur CBK	WrL	1996	N10 Brecon
W	Pod		BA Tender	BAT	19	08 Morriston, Swansea
W	Pod		Canteen Van	CaV	19	08 Morriston, Swansea
W	Pod		Chemical Incident Unit	CIU	19	08 Morriston, Swansea
W	Pod		Control Unit	CU	19	08 Morriston, Swansea
W	Pod		Flat Bed Lorry	FBL	19	08 Morriston, Swansea
W	Pod	Powell Duffryn Rolonoff	Foam Tender	FoT	19	08 Morriston, Swansea
W	Pod	HCB-Angus	Foam Tender	FoT	1993	08 Morriston, Swansea
W	Pod		Hose Laying Lorry	HL	19	08 Morriston, Swansea
W	Pod		Incident Control Unit	ICU	19	08 Morriston, Swansea
W	Pod	Leicester Carriage Co	Water Carrier	WrC	19	08 Morriston, Swansea

Notes: AFO144B & DFO855G were new to the Brecon & Radnor Joint Fire Brigade. EGP136J was new to London Transport. XUD565V, TJO728X and B645VUD were new to Oxfordshire Fire Service. ONP785Y was new to Hereford & Worcester Fire Brigade.

NORFOLK FIRE SERVICE

Norfolk County Fire Brigade, Hethersett, Norwich NR9 3DN

FNG779K	ERF 84PSA	HCB-Angus/Simon Snorkel SS70	HP	1972	47 Kings Lynn
NPW29P	Dodge K850	/	RT	1976	44 Fakenham
UCL493W	Dennis RS130	Dennis	WrT	1981	32 Bowthorpe
UCL495W	Dennis RS133	Dennis	WrT	1981	Reserve
XCL21X	Dennis Delta F125	Dennis/Simon Snorkel SS263	HP	1982	27 Norwich City
GEX533Y	Bedford TK1260A	Carmichael	ET	1983	54 Thetford
A51HNG	Dennis SS135	Dennis	WrT	1983	Reserve
A52HNG	Dennis SS135	Dennis	WrT	1983	Reserve
A53HNG	Dennis SS135	Dennis	WrT	1983	Reserve
A54HNG	Dennis SS135	Dennis	WrL	1983	42 Dereham
A250NNG	Dennis SS135	Dennis	WrT	1984	Reserve
A251NNG	Dennis SS135	Dennis	WrL	1984	21 Diss
A252NNG	Dennis SS135	Dennis	WrL	1984	28 Reepham
A253NNG	Dennis SS135	Dennis	WrL	1984	66 Cromer
B957PNG	Dennis SS135	Dennis	WrT	1984	70 Holt
B958PNG	Dennis SS135	Dennis	WrT	1984	31 Wymondham
B959PNG	Dennis SS135	Dennis	WrL	1984	31 Wymondham
B960PNG	Dennis SS135	Dennis	WrL	1984	44 Fakeham
B961PNG	Dennis SS135	Dennis	WrL	1984	51 Sandringham
C837YVF	Freight Rover Sherpa 350	Freight Rover	CU	1985	31 Wymondham
C710DNG	Dodge G16C	Angloco	WrC	1986	29 Sprowston
D533GVG	Mercedes-Benz 1222F	Polyma Fire Protection	WrT	1986	40 Attleborough
D481LPW	Mercedes-Benz 1222F	Polyma Fire Protection	ET	1987	69 Great Yarmouth
E345FBY	Mercedes-Benz 1222F	Rosenbauer	WrL	1987	23 Hethersett
E991PEX	Mercedes-Benz 1222F	Polyma Fire Protection	WrL	1988	30 Wroxham
E992PEX	Mercedes-Benz 1222F	Polyma Fire Protection	WrT	1987	51 Sandringham
E993PEX	Mercedes-Benz 1222F	Polyma Fire Protection	WrT	1988	66 Cromer
E207VCL	Mercedes-Benz 1222F	Polyma Fire Protection	WrT	1988	22 Harleston

Based at Spowston is Norfolk Fire Services' M235BVF, a Mercedes-Benz 1124F with Saxon Sanbec bodywork. The vehicle is seen on active duty and typifies the latest deliveries of pumping appliances to the brigade. *Philip Stephenson*

F666AEX	Renault Master T35	Reliance/Norfolk FB	ST	1988	23 Hethersett
F985XPW	Mercedes-Benz 1222F	Polyma Fire Protection	WrT	1988	65 Acle
F986XPW	Mercedes-Benz 1222F	Polyma Fire Protection	WrL	1988	58 Wells
F987XPW	Mercedes-Benz 1222F	Polyma Fire Protection	WrT	1988	21 Diss
F988XPW	Mercedes-Benz 1222F	Polyma Fire Protection	WrL	1988	46 Hunstanton
F516DVF	Mercedes-Benz 1222F	Polyma Fire Protection	WrL	1989	73 North Walsham
F517DVF	Mercedes-Benz 1222F	Polyma Fire Protection	WrT	1989	24 Hingham
F518DVF	Mercedes-Benz 1222F	Polyma Fire Protection	WrL	1989	20 Aylsham
F519DVF	Mercedes-Benz 1222F	Polyma Fire Protection	WrL	1989	41 Downham Market
F520DVF	Mercedes-Benz 1222F	Polyma Fire Protection	WrT	1989	52 Swaffham
G174JCL	Mercedes-Benz 1222F		WrT	1989	26 Long Stratton
G35KVG	Mercedes-Benz 1625/52	Mountain Range/Simon SS263	HP	1990	69 Great Yarmouth
H453SCL	Mercedes-Benz 1222F	Halton	WrT	1991	72 Mundesley
H454SCL	Mercedes-Benz 1222F	Halton	WrT	1991	48 Massingham
H455SCL	Mercedes-Benz 1222F	Halton	WrL	1991	74 Sheringham
H456SCL	Mercedes-Benz 1222F	Halton	WrT	1991	57 Watton
H457SCL	Mercedes-Benz 1222F	Halton	WrT	1991	70 Holt
H458SCL	Mercedes-Benz 1222F	Halton	WrT	1991	44 Fakenham
H553TVG	Mercedes-Benz 1222F	Halton	WrT	1991	45 Heacham
H554TVG	Mercedes-Benz 1222F	Halton	WrL	1991	Reserve
H556TVG	Mercedes-Benz 1222F	Halton	WrT	1991	50 Outwell
H557TVG	Mercedes-Benz 1222F	Halton	WrT	1991	71 Martham
H558TVG	Mercedes-Benz 1222F	Halton	WrT	1991	75 Stralham
H559TVG	Mercedes-Benz 1222F	Halton	WrT	1991	42 Dereham
J826CVF	Mercedes-Benz 1222F	Halton	WrT	1992	59 West Ealton
J827CVF	Mercedes-Benz 1222F	Halton	ET	1992	27 Norwich City
J828CVF	Mercedes-Benz 1222F	Halton	WrT	1992	53 Terrington St Clement
J829CVF	Mercedes-Benz 1222F	Halton	WrL	1992	28 Reepham
J830CVF	Mercedes-Benz 1222F	Halton	ET	1992	47 Kings Lynn
J831CVF	Mercedes-Benz 1222F	Halton	WrT	1992	49 Metheald
K643DEW	Renault Master T35	Wests/Norfolk FS	BACV	1993	Headquarters
L164RVG	Mercedes-Benz 1124F	Saxon Sanbec	WrL	1993	25 Loddon
L165RVG	Mercedes-Benz 1124F	Saxon Sanbec	WrL	1994	43 East harling
L166RVG	Mercedes-Benz 1124F	Saxon Sanbec	WrL	1994	Reserve
L167RVG	Mercedes-Benz 1124F	Saxon Sanbec	WrT	1994	Reserve
L168RVG	Mercedes-Benz 1124F	Saxon Sanbec	WrL	1994	69 Great Yarmouth
L726TNG	Mercedes-Benz 1124F	Saxon Sanbec	WrT	1994	54 Thetford
L727TNG	Mercedes-Benz 1124F	Saxon Sanbec	WrT	1994	67 Gorleston
L728TNG	Mercedes-Benz 1124F	Saxon Sanbec	WrL	1994	67 Gorleston
L729TNG	Mercedes-Benz 1124F	Saxon Sanbec	WrL	1994	54 Thetford
L730TNG	Mercedes-Benz 1124F	Saxon Sanbec	WrT	1994	47 Kings Lynn
M233BVF	Mercedes-Benz 1124F	Saxon Sanbec	WrL	1995	69 Great Yarmouth
M234BVF	Mercedes-Benz 1124F	Saxon Sanbec	WrL	1995	29 Spowston
M235BVF	Mercedes-Benz 1124F	Saxon Sanbec	WrL	1995	27 Norwich City
M236BVF	Mercedes-Benz 1124F	Saxon Sanbec	WrL	1995	27 Norwich City
M237BVF	Mercedes-Benz 1124F	Saxon Sanbec	WrL	1995	47 Kings Lynn
N284HEX	Mercedes-Benz 814D	Saxon Sanbec	CIU	1995	29 Spowston
trailer		400-litre Foam Tanker	FoT	1995	23 Hethersett
trailer		400-litre Foam Tanker	FoT	1995	47 Kings Lynn
trailer		Foam Container	FoT	1995	54 Thetford

Most of Norfolks pumping appliances are of Dennis or Mercedes-Benz manufacture. D533GVG represents the Mercedes-Benz 1222 type bodied here by Polyma Fire Protection. This WrT is located at Attleborough.
Edmund Gray

Two views illustrate the old and new order of Northamptonshire's pumping appliances. The top picture shows B457FVV, a Bedford KG with Saxon cab and bodywork which is now in the reserve fleet. The centre picture is L106VNV, an Excalibur-bodied Mercedes-Benz 1324 model that is currently on the run at Corby fire station. *Clive Shearman*

An uncommon use for a Scania chassis is that of a Water Carrier, Northamptonshire Fire and Rescue Service have L47JVV. On the run a WrC at Daventry. A P93M-250 chassis was used as the basis for this Maidment Tankers built appliance. *Robert Smith*

NORTHAMPTONSHIRE FIRE & RESCUE SERVICE

Northampton County Fire Brigade, Moulton Way, Moulton Park, Northampton NN3 1XJ

B457FVV	Bedford KG	Saxon Sanbec	WrT	1984	Reserve
B458FVV	Bedford KG	Saxon Sanbec	WrT	1984	Training Centre
C230MBD	Bedford TL750	Harrops	Cav	1985	22 Rushden
C231MBD	Bedford TL750	Harrops/NF&RS	OSU	1985	20 Wellingborough
D769YNH	Renault-Dodge G13	Alexander	WrT	1986	Reserve
D770YNH	Renault-Dodge G13	Alexander	WrT	1986	Reserve
D771YNH	Renault-Dodge G13	Alexander	WrT	1986	Reserve
E729JVV	Renault-Dodge G13	Alexander	WrT	1987	Reserve
E730JVV	Renault-Dodge G13	Alexander	WrL	1987	02 Towcester
E731JVV	Renault-Dodge G13	Alexander	WrL	1987	08 Brackley
F109CBD	Renault-Dodge G13	Alexander	WrT	1989	22 Rushden
F110CBD	Renault-Dodge G13	Alexander	WrT	1989	17 Burton Latimer
G681GRP	Renault-Dodge G13	Alexander	WrL	1990	09 Moulton
G682GRP	Renault-Dodge G13	Alexander	WrL	1990	11 Kettering
G683GRP	Renault-Dodge G13	Alexander	WrL	1990	20 Wellingborough
H581CNH	Dennis SS237	HCB-Angus	WrL	1990	16 Raunds
H582CNH	Dennis SS237	HCB-Angus	WrT	1990	13 Rothwell
H583CNH	Dennis SS237	HCB-Angus	WrL	1990	03 Woodford Halse
H943ENH	Dennis SS237	HCB-Angus	WrL	1991	04 Daventry
H105UBD	Scania P93M-280	Angloco/Bronto Skylift 28-2Ti	HP	1991	01 The Mounts, Northampton
H351VVV	Mercedes-Benz 1625	Rosenbauer/HIAB 140	ET	1991	10 Mereway, Northampton
J855PVV	Dennis RS237	Excalibur CBK	WrL	1992	21 Irthlingborough
J856PVV	Dennis RS237	Excalibur CBK	WrL	1992	14 Oundle
K45AVV	Mercedes-Benz 1726AK	W.H.Bence	HL	1992	09 Moulton
K141WVV	Dennis RS237	Excalibur CBK	WrL	1992	06 Guilsborough
K142WVV	Dennis RS237	Excalibur CBK	WrL	1992	19 Earls Barton
K143WVV	Dennis RS237	Excalibur CBK	WrL	1992	20 Wellingborough
K144WVV	Dennis RS237	Excalibur CBK	WrL	1992	15 Thrapston
L46JVV	Scania P93M-250	Angloco/Metz DLK30PLC	TL	1993	11 Kettering
L47JVV	Scania P93M-250	Maidment Tankers	WrC	1993	04 Daventry
L101JRP	Dennis RS237	Excalibur CBK	WrL	1993	05 Long Buckby
L102JRP	Dennis RS237	Excalibur CBK	WrL	1993	18 Desborough
L103JRP	Dennis RS237	Excalibur CBK	WrL	1993	04 Daventry
L104JRP	Dennis RS237	Excalibur CBK	WrL	1993	07 Brixworth
L105JRP	Dennis RS237	Excalibur CBK	WrL	1993	22 Rushden
L106VNV	Mercedes-Benz 1324	Excalibur CBK/Saxon Sanbec	WrL	1994	12 Corby
L107VNV	Mercedes-Benz 1324	Excalibur CBK/Saxon Sanbec	WrL	1994	11 Kettering
L108VNV	Mercedes-Benz 1324	Excalibur CBK/Saxon Sanbec	WrL	1994	01 The Mounts, Northampton
L109VNV	Mercedes-Benz 1324	Excalibur CBK/Saxon Sanbec	WrL	1994	09 Moulton
L110VNV	Mercedes-Benz 1324	Excalibur CBK/Saxon Sanbec	WrL	1994	10 Mereway, Northampton
L112VRP	Mercedes-Benz 1324	Excalibur CBK/Saxon Sanbec	WrL	1994	12 Corby

One water-carrier will be supplied during 1997.

NORTHUMBERLAND FIRE AND RESCUE SERVICE

Northumberland County Fire And Rescue Service, Loansdean, Morpeth NE61 2ED

T9	HTY91N	Ford D1114	North Eastern/Northumberland FS	WrL	1975	D19 Holy Island
T17	TFN697S	Ford D1114	North Eastern/Northumberland FS	WrL	1978	Reserve
T18	UVK225T	Ford D1114	North Eastern/Northumberland FS	WrL	1978	C09 Haydon Bridge
T22	BCU260V	Ford D1114	Angloco	WrL	1980	Training Centre
T23	FTN569W	Ford D1317	Angloco	WrL/ET	1981	Reserve
T24	FTN570W	Ford D1317	Angloco	WrL/ET	1981	Reserve
T25	LRG556X	Dodge G1313	Carmichael	WrL/ET	1981	D17 Alnwick
T26	LRG557X	Dodge G1313	Carmichael	WrL/ET	1981	D14 Berwick on Tweed
T27	LRG555X	Dodge G1313	Carmichael	WrL/ET	1981	A01 Morpeth
T28	LRG558X	Dodge G1313	Carmichael	WrL/ET	1981	D16 Seahouses
T29	F32MNL	Volvo FL6.14	Angloco	WrL/ET	1989	C10 Haltwhistle
T30	F995NRG	Volvo FL6.14	Angloco	WrL/ET	1989	D13 Wooler
T32	G458UCN	Volvo FL6.14	Angloco	WrL/ET	1990	C11 Bellingham
T33	G189UTN	Volvo FL6.14	Angloco	WrL/ET	1990	C07 Hexham
T34	G421XCN	Volvo FL6.14	Fulton & Wylie	WrL/ET	1990	C08 Allendale
T35	G678XCN	Volvo FL6.14	Fulton & Wylie	WrL/ET	1990	D15 Belford
T36	H401EFT	Volvo FL6.14	Fulton & Wylie	WrL/ET	1991	D17 Alnwick
T37	H402EFT	Volvo FL6.14	Fulton & Wylie	WrL/ET	1991	A12 Rothbury
T38	J459MCU	Volvo FL6.14	Fulton & Wylie	WrL/ET	1992	B06 Prudhoe
T39	J460MCU	Volvo FL6.14	Fulton & Wylie	WrL/ET	1992	A18 Amble
T40	K158TBB	Volvo FL6.14	Carmichael International	WrL/ET	1993	B03 Blyth
T41	K162TBB	Volvo FL6.14	Carmichael International	WrL/ET	1993	B05 Ponteland
T42	M235KTN	Volvo FL6.14	Carmichael International	WrL/ET	1995	A01 Morpeth
T43	M236KTN	Volvo FL6.14	Carmichael International	WrL/ET	1995	B03 Blyth
T44	N336ORG	Volvo FL6.14	Carmichael International	WrL/ET	1995	A02 Ashington
T45	N47OTN	Volvo FL6.14	Carmichael International	WrL/ET	1995	C07 Hexham
T46	N794PBB	Volvo FL6.14	Carmichael International	WrL/ET	1995	D14 Berwick on Tweed
T47	N795PBB	Volvo FL6.14	Carmichael International	WrL/ET	1995	B04 Cramlington
T48	P151XBB	Volvo FL6.14	Carmichael International	WrL/ET	1996	
T49	P152XBB	Volvo FL6.14	Carmichael International	WrL/ET	1996	
T50	P875XCN	Volvo FL6.14	Carmichael International	WrL/ET	1996	
T51	P876XCN	Volvo FL6.14	Carmichael International	WrL/ET	1996	
U6	G477WVK	Land Rover 90 Tdi	Land Rover/Northumberland FS	L4V	1990	D14 Berwick on Tweed
U7	D496YNL	Peugeot 504	Peugeot Pick Up	L2V	1987	A02 Ashington
U21	H654ECU	Volvo FL6.17	Penman Multilift	PM	1991	Training Centre
U22	H508EFT	Land Rover 90 Defender Tdi	Land Rover/Northumberland FS	L4V	1991	C07 Hexham
U24	J119LTY	Renault Master T35	Crystals	PCV	1992	Headquarters
U26	K499PFT	Mercedes-Benz 310D	Northumberland FS	OSU/GPV	1992	Stores
U28	K318SNL	Land Rover Discovery Tdi	Northumberland FS	CU/L4V	1992	Headquarters
U31	L566KKS	Iveco Daily 49.12		CFS	1994	Headquarters
U36	N337ORG	Vauxhall Brava 4x4	Vauxhall	L4V	1995	A01 Morpeth
	unreg	Kawasaki Bayou 4x4 & Trailer	Northumberland FS	ATV	1990	Headquarters
	G87SNL	Iveco TurboZeta	Northum FS	ISU	1990	Workshops
M1	Trailer	Durham Trailers	Recovery Unit	Rec	1991	A01 Morpeth
M2	Trailer	Tow-a-Van	Lighting and pump Unit	LPU	1988	D14 Berwick on Tweed
M3	Pod	Penman	BA Training Unit	BA(T)	1991	Headquarters
M4	Pod	Penman	Incident Command & Control Unit	ICCU	1991	Headquarters
M5	Pod	C&S Engineering	Flat-bed lorry	FBL	1991	Headquarters
M6	Trailer	Tow-a-Van	Chemical Incident Unit	CIU	1994	A01 Morpeth
M7	Trailer	Tow-a-Van	Community Fire Safety	CFS	1994	Headquarters
M8	Trailer	Tow-a-Van	Lighting and Pump Unit	LPU	1991	B04 Cramlington
M9	Trailer	Tow-a-Van	Lighting and Pump Unit	LPU	1991	C07 Hexham
M10	Trailer	Tow-a-Van	Lighting and Pump Unit	LPU	1995	?
M11	Trailer	Tow-a-Van	Chemical Incident Unit	CIU	1995	C07 Hexham
M12	Trailer	Tow-a-Van	Fire Exhibition Unit	ExU	1995	Headquarters
M13	Trailer	Tow-a-Van	Lighting and Pump Unit	LPU	1996	A01 Morpeth

Opposite: **Northumbria Fire & Rescue Services' Ford pumping appliances are gradually being phased out by Volvo-based products. Shown here are Volvo FL6, G421XCN, with bodywork in this example by Fulton & Wylie, though later deliveries have Carmichael International bodywork, and HTY91N, a Ford D1114 with bodywork by North Eastern and Northumberland Fire Service.** *Alan Hewitt*

The Fire Brigade Handbook

NORTH WALES FIRE SERVICE

North Wales Fire Service, Coast Road, Rhyl, LL18 3PL

(C - acquired from County of Clwyd, FB, G - acquired from Gwynedd FS)

C	MEY839R	Land Rover Series III 109	HCB-Angus	L4P	1977	C08 Corwen
C	PEY254S	Dennis Delta 2 F125	Dennis/Simon Snorkel SS263	HP	1978	Reserve
C	YHO56V	Land Rover Series III 109	HCB-Angus	L4P	1979	C11 Llangollen
C	ATU291W	Dodge 50 S35C	Besco	CaV	1981	E02 Deeside
C	YFM675W	Dennis RS133	Dennis	WrL	1981	Driving School
C	DCC566X	Dodge G1313	Carmichael	WrL	1981	Reserve
G	DCC920X	Dodge G8685	Carmichael	PL	1981	Training School
G	DCC921X	Dodge G8685	Carmichael	PL	1981	Reserve
G	DCC922X	Dodge G8685	Carmichael	PL	1981	Reserve
G	DCC923X	Dodge G8685	Carmichael	PL	1981	Reserve
G	GCC334Y	Dodge G12C	Carmichael	WrL	1982	W21 Tywyn
G	GCC335Y	Dodge G12C	Carmichael	WrL	1982	W19 Pwllheli
G	A914KJC	Dodge G16C	Carmichael/Magirus DL30U	TL	1984	C01 Llandudno
C	A758LJC	Dennis RS135	Dennis	WrL	1984	E02 Deeside
G	B508MJC	Dodge G12C	Carmichael	WrL	1984	W10 Beaumaris
C	B284OEY	Dodge 50 S66C	Fulton & Wylie	ET	1985	Reserve
C	B36OJC	Dodge G13C	Fulton & Wylie	WrL	1985	Reserve
C	B37OJC	Dodge G13C	Fulton & Wylie	WrL	1985	Reserve
C	Trailer	Portakabin 1980/Lynton	Chemical Incident Unit	CIU	1985	C02 Colwyn Bay
C	Trailer	Portakabin 1981/Lynton	Control Unit	CU	1985	C02 Colwyn Bay
C	C477REY	Bedford TL750 Articulated	Lynton Commercials	TU	1985	C02 Colwyn Bay
C	C478REY	Bedford TL750 Articulated	Lynton Commercials	TU	1985	C02 Colwyn Bay
C	D999TFY	Volvo FL6.16	Fulton & Wylie	WrL	1987	C15 St. Asaph
G	D884UCC	Renault-Dodge G12C	Carmichael	WrL	1987	W05 Abersoch
G	D885UCC	Renault-Dodge G12C	Carmichael	WrL	1987	C05 Betws-Y-Coed
G	D886UCC	Renault-Dodge G12C	Carmichael	WrL	1987	W06 Aberdovey
C	E499UTJ	Volvo FL6.16	Fulton & Wylie	WrL	1987	C09 Denbigh
G	E172XCC	Renault-Dodge G12C	Carmichael	WrL	1987	W12 Blaenau Ffestiniog
G	E173XCC	Renault-Dodge G12C	Carmichael	WrL	1987	C12 Llanrwst

Among the last new appliances were three Mercedes-Benz appliances, two WrL's and an ET. Seen here is the ET (M53BEY) with an 1124AF chassis, Saxon Sanbec Crewcab, Angloco coachwork and an HIAB 090 crane.
Angloco Ltd

C	E799VWM	Volvo FL6.14	Fulton & Wylie	WrL	1988	E04 Chirk
C	E999WHF	Volvo FL6.14	Fulton & Wylie	WrL	1988	E05 Flint
G	E199XJC	Renault-Dodge G16C	Carmichael/Magirus DL30	TL	1988	W02 Bangor
G	E542YEY	Land Rover 110 V8	Carmichael	L4P	1988	W18 Porthmadog
G	E543YEY	Land Rover 110 V8	Carmichael	L4P	1988	W01 Caernarfon
G	E544YEY	Land Rover 110 V8	Carmichael	L4P	1988	W19 Pwllheli
G	E545YEY	Land Rover 110 V8	Carmichael	L4P	1988	C12 Llanrwst
G	E546YEY	Land Rover 110 V8	Carmichael	L4P	1988	C05 Betws-Y-Coed
C	F276KLG	Land Rover 110 Tdi	Saxon Sanbec	L4P	1988	E04 Chirk
G	F277PFF	Renault-Dodge G12C	Carmichael	WrL	1988	W17 Nefyn
G	F278PFF	Renault-Dodge G12C	Carmichael	WrL	1988	W08 Bala
G	F279PFF	Renault-Dodge G12C	Carmichael	WrL	1988	W19 Pwllheli
G	F280PFF	Renault-Dodge G12C	Carmichael/Gwynedd FS	ET	1989	W01 Caernarfon
C	F166AFY	Volvo FL6.14	Fulton & Wylie	WrL	1989	E01 Wrexham
C	F888AKC	Volvo FL6.14	Fulton & Wylie	WrL	1989	E08 Mold
C	Trailer	Lynton Commercials	Fire Prevention Unit	FPU	1989	C03 Rhyl
C	G896DKD	Volvo FL6.14	Fulton & Wylie	WrL	1989	C11 Llangollen
C	G897DKD	Volvo FL6.14	Fulton & Wylie	WrL	1990	E06 Holywell
C	G899DKD	Volvo FL6.14	Fulton & Wylie	WrL	1989	C14 Ruthin
G	G245GEY	Renault-Dodge G13C	Carmichael	WrL	1989	W14 Llanberis
G	G246GEY	Renault-Dodge G13C	Carmichael	WrL	1989	W15 Llangefni
G	G247GEY	Renault-Dodge G13C	Carmichael	WrL	1989	W07 Amlwch
G	G248GEY	Renault-Dodge G13C	Carmichael	WrL	1989	W18 Porthmadog
G	G479GJC	Land Rover 110 V8	Carmichael	L4P	1989	W21 Tywyn
G	G480GJC	Land Rover 110 V8	Carmichael	L4P	1989	W11 Benllech
G	H734LEY	Mercedes-Benz 814D	Carmichael	ET	1990	W04 Dolgellau
G	H721LJC	Land Rover Defender 110 V8	Carmichael	L4P	1990	C07 Conwy
G	H722LJC	Land Rover Defender 110 V8	Carmichael	L4P	1991	W12 Blaenau Ffestiniog
C	H72HKC	Volvo FL6.14	Fulton & Wylie	WrL	1991	C13 Prestatyn
C	H73HKC	Volvo FL6.14	Fulton & Wylie	WrL	1991	E03 Buckley
C	H74HKC	Volvo FL6.14	Fulton & Wylie	WrL	1991	C02 Colwyn Bay
G	H762MCC	Mercedes-Benz 1120F	Carmichael	WrL	1991	W20 Rhosneigr
G	H763MCC	Mercedes-Benz 1120F	Carmichael	WrL	1991	W16 Menai Bridge
G	H764MCC	Mercedes-Benz 1120F	Carmichael	WrL	1991	W11 Benllech
C	J681FLG	Land Rover Defender 110 Tdi	Saxon Sanbec	L4P	1991	C04 Abergele
G	J473PCC	Land Rover Defender 110 V8	Carmichael	L4P	1992	W09 Barmouth
G	J474PCC	Mercedes-Benz 1120F	Carmichael	WrL	1992	W13 Harlech
G	J475PCC	Mercedes-Benz 1120F	Carmichael	WrL	1992	W03 Holyhead
G	J476PCC	Mercedes-Benz 1120F	Carmichael	WrL	1992	C10 Llanfairfechan
C	J21PJC	Scania P113M-280	Angloco/Bronto Skylift 28-2Ti	ALP	1992	E01 Wrexham
C	K83SCC	Mercedes-Benz 1124F	Rosenbauer/Angloco	WrL	1992	C04 Abergele
C	K84SCC	Mercedes-Benz 1124F	Rosenbauer/Angloco	WrL	1992	C06 Cerrig-Y-Drudion
C	K85SCC	Mercedes-Benz 1124F	Rosenbauer/Angloco	WrL	1992	C03 Rhyl
G	K89SCC	Mercedes-Benz 609D	Saxon Sanbec	CU	1992	W01 Caernarfon
G	K183SCC	Ford Transit	Saxon Sanbec	L2P	1992	W13 Harlech
C	K31TCC	Mercedes-Benz 1124F	Saxon Sanbec	WrL	1992	E07 Johnstown
C	K32TCC	Mercedes-Benz 1124F	Saxon Sanbec	WrL	1993	C08 Corwen
C	K34TCC	Mercedes-Benz 1124F	Saxon Sanbec	WrL	1993	E01 Wrexham
G	K275TCC	Mercedes-Benz 1124F	Saxon Sanbec	WrL	1993	W04 Dolgellau
G	L247VJC	Ford Transit	Saxon Sanbec	L2P	1993	W03 Holyhead
G	L482WCC	Mercedes-Benz 1124F	Saxon Sanbec	WrL	1993	W02 Bangor
G	L483WCC	Mercedes-Benz 1124F	Saxon Sanbec	WrL	1993	W09 Barmouth
G	L484WCC	Mercedes-Benz 1124F	Saxon Sanbec	WrL	1993	C07 Conwy
C	L21XEY	Volvo FL10	Angloco/Bronto Skylift F30HDT	ALP	1994	C03 Rhyl
G	M525AEY	Mercedes-Benz 1124F	Saxon Sanbec	WrL	1994	C01 Llandudno
G	M526AEY	Mercedes-Benz 1124F	Saxon Sanbec	WrL	1994	W03 Holyhead
G	M527AEY	Mercedes-Benz 1124F	Saxon Sanbec	WrL	1994	W01 Caernarfon
C	M51ACC	Mercedes-Benz 1124F	Angloco	WrL	1995	E01 Wrexham
C	M52ACC	Mercedes-Benz 1124F	Angloco	WrL	1995	E02 Deeside
C	M51BEY	Mercedes-Benz 1124F	Saxon Sanbec	WrL	1995	C02 Colwyn Bay
C	M52BEY	Mercedes-Benz 1124F	Saxon Sanbec	WrL	1995	C03 Rhyl
C	M53BEY	Mercedes-Benz 1124AF	Saxon Sanbec/Angloco/HIAB90	ET	1995	E01 Wrexham
C	N21DJC	Renault B120-60/R Boughton	Bedwas	L4P	1996	C14 Ruthin
C	N23DJC	Mercedes-Benz 1124AF	Saxon Sanbec/Angloco/HIAB90	ET	1996	C02 Colwyn Bay
G	N230ECC	Mercedes-Benz 1124F	Saxon Sanbec	WrL	1996	W02 Bangor
G	N231ECC	Mercedes-Benz 1124F	Saxon Sanbec	WrL	1996	C01 Llandudno
G	N232ECC	Mercedes-Benz 1124F	Saxon Sanbec	WrL	1996	W01 Caernarfon
G	N662ECC	Ford Transit/County 4x4	Saxon Sanbec	L4P	1996	W07 Amlwch
G	N663ECC	Ford Transit	Saxon Sanbec	L2P	1996	W15 Llangefni
G	N664ECC	Ford Transit/County 4x4	Saxon Sanbec	L4P	1996	W04 Dolgellau

NORTH YORKSHIRE FIRE & RESCUE SERVICE

North Yorkshire Fire & Rescue Service, Crosby Road, Northallerton   DL6 1AB

Reg	Chassis	Body	Type	Year	Location
PAJ576R	ERF 84PS	Metz/Metz DL30	TL	1976	Reserve
DPY593V	Bedford TK(SWB)	HCB-Angus	WrL	1979	Reserve
GHN545V	Bedford TK1260	Ray Smith	PM	1980	W2 Ripon
KVN296W	Bedford TK/USG 4x4	HCB-Angus	WrL	1980	E8 Danby
Pod	Ray Smith/Cocker	Control Unit	CU	1980	W2 Ripon
Pod	Ray Smith/Cocker	Canteen Van	CaV	1980	S6 Tadcaster
NDC406W	Bedford TK1260	Wallace Arnold/Ray Smith	PM	1981	S6 Tadcaster
OPY790X	Bedford TKG	Carmichael	WrL	1981	W6 Skipton
OPY791X	Bedford TKG	Carmichael	WrL	1981	Reserve
OPY792X	Bedford TKG	Carmichael	WrL	1981	Training School
OPY793X	Bedford TKG	Carmichael	WrL	1981	Training School
Pod	Ray Smith/W.H.Hull	Flat Bed Lorry/Crane	FBL	1981	E11 Northallerton
Pod	Ray Smith/Cocker	Fire Prevention Unit	FPU	1981	E11 Northallerton
YVN68Y	Bedford TKG	Carmichael	WrL	1982	E11 Northallerton
YVN69Y	Bedford TKG	Carmichael	WrL	1982	Reserve
A78GVN	Bedford TKG	Carmichael	WrL	1983	Reserve
A79GVN	Bedford TKG	Carmichael	WrL	1983	S7 Selby
A81GVN	Bedford TKG	Carmichael	WrL	1983	Reserve
A82GVN	Bedford TKG	Carmichael	WrL	1983	S2 Acomb, York
A262JAJ	Bedford TK1260	Vincent Greenhous/Ray Smith	PM	1983	S6 Tadcaster
Pod	Ray Smith/Clayton Tankers	Water Carrier	WrC	1983	E11 Northallerton
Pod	Ray Smith/Burton & Greaves	Foam/Salvage Tender	FoST	1983	S6 Tadcaster
B179NHN	Bedford TKG	Carmichael	WrL	1984	Reserve
B180NHN	Bedford TKG	Carmichael	WrL	1984	Reserve
B181NHN	Bedford TKG	Carmichael	WrL	1984	Reserve
B182NHN	Bedford TKG	Carmichael	WrL	1984	Reserve

Only 7 French built Camiva turntable ladders operate with UK Fire Brigades.   K667BPY is the only one with a Scania P93ML chassis and this carries Bedwas bodywork. The appliance is seen at York and shows the tapering ladder common to all these British registered Camiva TLs. *Tony Wilson*

C923VHN	Bedford TKG	Carmichael	WrL	1985	W3 Masham
C924VHN	Bedford TKG	Carmichael	WrL	1985	Reserve
C925VHN	Bedford TKG	Carmichael	WrL	1985	S3 Boroughbridge
C926VHN	Bedford TKG	Carmichael	WrL	1985	W2 Ripon
C927VHN	Bedford TKG	Carmichael	WrL	1985	E2 Malton
C112WEF	Land Rover 110	North Yorkshire F&RS	L4P	1985	W14 Grassington Volunteers
D429EHN	Volvo FL6.17	Angloco/Metz DL30	TL	1986	E1 Scarborough
D843EPY	Bedford TL	Carmichael	WrL	1986	E14 Stokesley
D844EPY	Bedford TL	Carmichael	WrL	1986	S6 Tadcaster
D845EPY	Bedford TL	Carmichael	WrL	1986	S4 Easingwold
D830EVN	Bedford TL1260	Ray Smith	PM	1986	E11 Northallerton
D490GAJ	Crayford Engineering Argocat 8x8	Crayford Engineering	ATV	1986	E13 Kirkbymoorside
E103HEF	Land Rover County 110	Land Rover	L4V	1987	E1 Scarborough
E473KVN	Volvo FL6.14	Carmichael	WrL	1987	W13 Leyburn
E474KVN	Volvo FL6.14	Carmichael	WrL	1987	E6 Filey
E475KVN	Volvo FL6.14	Carmichael	WrL	1987	Driving School
E788LAJ	Dennis DS153	HCB-Angus	ET	1987	Training Centre
E789LAJ	Dennis DS153	HCB-Angus	ET	1987	W6 Skipton
E280NVN	Volvo FL6.17	Angloco/Metz DL30H	TL	1987	W1 Harrogate
F902CUA	Land Rover County 110	North Yorkshire F&RS	L4P	1988	E15 Goathland Volunteers
F536UAJ	Volvo FL6.14	Carmichael	WrL	1988	W11 Hawes
F537UAJ	Volvo FL6.14	Carmichael	WrL	1988	E3 Pickering
F538UAJ	Volvo FL6.14	Carmichael	WrL	1988	W6 Skipton
F539UAJ	Volvo FL6.14	Carmichael	WrL	1988	E12 Helmsley
F540UAJ	Volvo FL6.14	Carmichael	WrL	1988	S5 Thirsk
F541UAJ	Volvo FL6.14	Carmichael	WrL	1988	W10 Reeth
F304VDC	Dennis DS153	HCB-Angus	ET	1988	E1 Scarborough
G353BDC	Volvo FL6.14	Carmichael	WrL	1989	... Colburn
G354BDC	Volvo FL6.14	Carmichael	WrL	1989	W7 Bentham
G355BDC	Volvo FL6.14	Carmichael	WrL	1989	W5 Knaresborough
G356BDC	Volvo FL6.14	Carmichael	WrL	1989	W8 Settle
G357BDC	Volvo FL6.14	Carmichael	WrL	1989	E4 Snainton
G713DAJ	Dennis DS153	HCB-Angus	ET	1989	W1 Harrogate
H476JAJ	Volvo FL6.14	Carmichael	WrL	1990	E1 Scarborough
H477JAJ	Volvo FL6.14	Carmichael	WrL	1990	???
H478JAJ	Volvo FL6.14	Carmichael	WrL	1990	E5 Sherburn
H479JAJ	Volvo FL6.14	Carmichael	WrL	1990	W12 Bedale
H481JAJ	Volvo FL6.14	Carmichael	WrL	1990	W1 Harrogate
H551MAJ	Dennis DS153	HCB-Angus	ET	1990	S1 York
J572RPY	Volvo FL6.14	HCB-Angus	WrL	1992	E11 Northallerton
J573RPY	Volvo FL6.14	HCB-Angus	WrL	1992	W1 Harrogate
J574RPY	Volvo FL6.14	HCB-Angus	WrL	1992	E1 Scarborough
J575RPY	Volvo FL6.14	HCB-Angus	WrL	1992	S2 Acomb, York
J576RPY	Volvo FL6.14	HCB-Angus	WrL	1992	E9 Lythe
K125APY	Mercedes-Benz 917AF	HCB-Angus	S4x4	1992	E13 Kirbymoorside
K263YPY	Dennis DS153	HCB-Angus	ET	1992	E11 Northallerton
K667BPY	Scania P93ML-250	Bedwas/Camiva EPA30	TL	1993	S1 York
K816XPY	Volvo FL6.14	HCB-Angus	RP	1993	S6 Tadcaster
L795GVN	Mercedes-Benz 917AF	Angloco	S4x4	1993	E14 Stokesley
L796GVN	Mercedes-Benz 917AF	Angloco	S4x4	1993	W4 Summerbridge
L797GVN	Mercedes-Benz 917AF	Angloco	S4x4	1993	W13 Leyburn
L798GVN	Mercedes-Benz 917AF	Angloco	S4x4	1993	E10 Robin Hoods Bay
L799GVN	Mercedes-Benz 917AF	Angloco	S4x4	1993	S1 Thirsk
(9120)	ScotTrack Glencoe ATV	ScotTrack	ATV	1993	E13 Kirbymoorside
Boat	Humber Alpha	Rescue Boat & Trailer	IFB	1994	S1 York
L691GPY	Mercedes-Benz 817D	Bott/North Yorkshire F&RS	ISU	1994	S2 Acomb, York
Pod	Ray Smith/Aim Hydraulics	Flat Bed Lorry/Crane	FBL	1994	S6 Tadcaster
Pod	Bill Scott	Decontamination Unit	DecU	1994	S6 Tadcaster
L282NWY	Land Rover Defender 110 Tdi	Land Rover County	TU	1993	E11 Northallerton
M654RVN	Volvo FL6.14	Emergency One	RP	1995	E7 Whitby
M655RVN	Volvo FL6.14	Emergency One	RP	1995	W9 Richmond
M656RVN	Volvo FL6.14	Emergency One	RP	1995	S7 Selby
M657RVN	Volvo FL6.14	Emergency One	RP	1995	E2 Malton
(9121)	ScotTrack Glencoe ATV	ScotTrack	ATV	1995	E11 Northallerton
N820WPY	Volvo FL6.14	Emergency One	WrL	1996	S1 York
N821WPY	Volvo FL6.14	Emergency One	WrL	1996	S1 York
N822WPY	Volvo FL6.14	Emergency One	RP	1996	W2 Ripon
N765YEF	Volvo FL6.14	Baileys Totalfleet	ICU	1996	E11 Northallerton
P	Volvo FL6.14	Emergency One	RP	1996	
P	Volvo FL6.14	Emergency One	RP	1996	

NOTTINGHAMSHIRE FIRE & RESCUE SERVICE

Nottingham Fire & Rescue Service, Bestwood Lodge, Arnold, Nottingham NG5 8PD

YTO999K	AEC Mercury	HCB-Angus	PE	1971	Preserved at 20
RCH638L	Bristol VRT/SL2/6G	Eastern Coach Works	ExU	1973	Fire Safety Department
WNU799S	Dennis Delta 2 F125	Dennis/Simon Snorkel SS220	HP	1978	Reserve
KAL699V	Dennis RS133	Dennis	RT	1980	Reserve
KAL700V	Dennis RS133	Dennis	WrL/RT	1980	Training Centre
NTO886W	Bedford TL1470	Powell Duffryn Multilift	PM	1981	22 Beeston
NTO887W	Bedford TL1470	Powell Duffryn Multilift	PM	1981	20 Stockhill

There are many differences between the Dennis Sabre and the restyled Rapier appliances. Rapiers have chassis frames enabling a lower centre of gravity by virtue of the space frame construction method, while the Sabres have a more conventional method of construction resulting in a higher frame and a consequently taller appliance. Note how the Sabre has a deeper front bumper and additional access steps for the crew cab when compared with the Rapier.
Shown here are Nottinghamshire's M930RRB (Rapier) and N909VV0, (Sabre) with John Dennis bodywork.
Graham Brown

TTV956X	Dennis RS133	Dennis	WrL	1982	Reserve
TTV958X	Dennis RS133	Dennis	RT	1982	Reserve
WNU484Y	Bedford TL1470	Powell Duffryn Multilift	PM	1982	01 Mansfield
A52CTV	Shelvoke & Drewry WY	Angloco/Metz DL30	TL	1984	18 Central, Nottingham
B369HAL	Dennis RS135	Fulton & Wylie	WrL	1984	Reserve
B370HAL	Dennis RS135	Fulton & Wylie	WrL	1984	Training Centre
B371HAL	Dennis RS135	Dennis	WrL	1984	Training Centre
B941JRB	Dennis RS135	Dennis	WrL	1985	Reserve
B942JRB	Dennis RS135	Dennis	WrL	1985	Reserve
B943JRB	Dennis RS135	Dennis	WrL	1985	Reserve
B944JRB	Dennis RS135	Dennis	WrL	1985	02 Blidworth
C141PRA	Dennis RS135	Fulton & Wylie	WrL	1986	28 East Leake
C142PRA	Dennis RS135	Fulton & Wylie	WrL	1986	27 Carlton
C143PRA	Dennis RS135	Fulton & Wylie	WrL	1986	15 Collingham
C144PRA	Dennis RS135	Fulton & Wylie	WrL	1986	26 Arnold
C145PRA	Dennis RS135	Fulton & Wylie	WrL	1986	11 Misterton
E700CRC	Dennis RS135	Fulton & Wylie	WrL	1988	08 Worksop
E701CVO	Dennis RS135	Fulton & Wylie	WrL	1988	23 Stapleford
E702CVO	Dennis RS135	Fulton & Wylie	WrL	1988	14 Southwell
E703CVO	Dennis RS135	Fulton & Wylie	WrL	1988	05 Kirkby in Ashfield
F990GRC	Renault-Dodge G13	Carmichael	RT	1988	12 Retford
F999GRC	Renault-Dodge G13	Carmichael	RT	1988	16 Newark
F999HVO	Dennis RS135	Fulton & Wylie	WrL	1988	13 Tuxford
F997LAU	Dennis RS137	Fulton & Wylie	WrL	1989	06 Edwinstowe
F998LAU	Dennis RS137	Fulton & Wylie	WrL	1989	17 Bingham
F999LAU	Dennis RS137	Fulton & Wylie	WrL	1989	01 Mansfield
H994VTO	Dennis RS237	Fulton & Wylie	WrL	1990	22 Beeston
H995VTO	Dennis RS237	Fulton & Wylie	WrL	1990	16 Newark
H996VTO	Dennis RS237	Fulton & Wylie	WrL	1990	2 Budworth
H997VTO	Dennis RS237	Fulton & Wylie	WrL	1990	12 Retford
H998VTO	Dennis RS237	Fulton & Wylie	WrL	1990	07 Warsop
H299XRR	Volvo FL10	Saxon/Simon Snorkel/ST240-S	HP	1991	01 Mansfield
H449WCH	Leyland-DAF Roadrunner 10.15	Flat Bed Lorry	FBL	1990	Training Centre
J378BNU	Land Rover Defender 110 Tdi	Land Rover	PCV	1991	01 Mansfield
J381BNU	Land Rover Defender 110 Tdi	Land Rover	PCV	1991	27 Carlton
J350CNU	Iveco Turbo-Zeta 79.14	Mellor	BAU	1991	22 Beeston
K373FVO	Dennis RS237	John Dennis	WrL	1992	10 Harworth
L445MNN	Renault Midliner M230	Carmichael International	RT	1993	05 Kirkby in Ashfield
L391MRB	Dennis RS237	John Dennis	WrL	1993	25 Hucknall
L392MRB	Dennis RS237	John Dennis	WrL	1993	24 Eastwood
L393MRB	Dennis RS237	John Dennis	WrL	1993	Training Centre
L538MRB	Ford Transit	Ford Minibus	PCV	1994	Training Centre
M920RRB	Dennis Rapier TF203	John Dennis	WrL	1994	18 Central, Nottingham
M930RRB	Dennis Rapier TF203	John Dennis	WrL	1994	18 Central, Nottingham
M940RRB	Dennis Rapier TF203	John Dennis	WrL	1994	08 Worksop
M950RRB	Dennis Rapier TF203	John Dennis	WrL	1994	05 Kirkby in Ashfield
N901NVO	Dennis Rapier R411	John Dennis	WrL	1996	01 Mansfield
N902NVO	Dennis Rapier R411	John Dennis	WrL	1996	19 West Bridgford
N903NVO	Dennis Rapier R411	John Dennis	WrL	1996	19 West Bridgford
N908NVO	Dennis Sabre S411	John Dennis	WrL	1995	12 Reford
N909NVO	Dennis Sabre S411	John Dennis	WrL	1995	26 Arnold
N910NVO	Dennis Sabre S411	John Dennis	WrL	1995	27 Carlton
N633XTO	Dennis Sabre S411	John Dennis	WrL	1996	22 Beeston
N634XTO	Dennis Sabre S411	John Dennis	WrL	1996	20 Stockhill
N635XTO	Dennis Rapier R411	John Dennis	WrL	1996	20 Stockhill
N650XTO	Dennis Rapier R411	John Dennis	WrL	1996	16 Newark
N618YRR	Ford Transit VE6	Ford	PCV	1996	18 Central, Nottingham
P418CAL	Volvo FL6-18	Angloco/Metz	TL	1996	18 Central, Nottingham

GPL1	Pod	Neville EMV	General Purpose Vehicle	GPV	1982	01 Mansfield
GPL2	Pod	Neville EMV	General Purpose Vehicle	GPV	1982	08 Worksop
	Pod	Neville EMV	Foam Tender	FOT	1982	01 Mansfield
	Pod	Neville EMV	Hose Laying Lorry	HL	1982	01 Mansfield
	Pod	Neville EMV	Control Unit	CU	1982	01 Mansfield

Note: - RCH638L was originally a bus new to Trent Motor Traction Ltd.

OXFORDSHIRE FIRE SERVICE

Oxfordshire Fire Service, Sterling Road, Kidlington OX5 2DU

Pod	Landsman	Foam Carrier	FoT	1976	A6 Kidlington
CFC601Y	Bedford TL1630	Carmichael/Magirus DL30U	TL	1981	B1 Rewley Road, Oxford
D812OBW	Bedford TL1260	HCB-Angus	WrL	1986	Reserve
D813OBW	Bedford TL1260	HCB-Angus	WrL	1986	Reserve
D814OBW	Bedford TL1200	Multilift	PM	1986	A6 Kidlington
D815OBW	Bedford TL1200	Multilift	PM	1986	B1 Rewley Road, Oxford
D816OBW	Bedford TL1200	Multilift	PM	1986	A6 Kidlington
Pod	Wilsden	Communications Unit	CU	1988	A6 Kidlington
E570AJO	Volvo FL6.14	HCB-Angus	WrL	1988	Reserve
E571AJO	Volvo FL6.14	HCB-Angus	WrL	1988	Reserve
E572AJO	Volvo FL6.14	HCB-Angus	WrL	1988	Training School
E573AJO	Volvo FL6.14	HCB-Angus	WrL	1988	Training School
E902WUD	Volvo FL6.17	Eagle/Frazer/Oxfordshire FS	WrC	1988	B10 The Slade, Oxford
F904HWL	Volvo FL6.14	Excalibur CBK	WrL	1989	A8 Deddington
F905HWL	Volvo FL6.14	Excalibur CBK	WrL	1989	A5 Woodstock
F906HWL	Volvo FL6.14	Excalibur CBK	WrL	1989	Reserve
F907HWL	Volvo FL6.14	Excalibur CBK	WrL	1989	Reserve
F933HFC	Land Rover 90 Tdi	Land Rover	L4V	1989	Headquarters
G220KLS	Leyland	Tanker	WC	1990	A9 Eynsham
G181SJO	Volvo FL6.14	Excalibur CBK	WrL	1990	B3 Faringdon
G182SJO	Volvo FL6.14	Excalibur CBK	WrL	1990	B3 Faringdon
G183SJO	Volvo FL6.14	Excalibur CBK	WrL	1990	B7 Thame
G184SJO	Volvo FL6.14	Excalibur CBK	WrL	1990	B7 Thame
G775UBW	Volvo FL10	Angloco/Bronto Skylift 28-2Ti	HP	1990	B1 Rewley Road, Oxford
H421FUD	Volvo FL6.14	Excalibur CBK	WrL	1991	A7 Bicester
H422FUD	Volvo FL6.14	Excalibur CBK	WrL	1991	A7 Bicester
H423FUD	Volvo FL6.14	Excalibur CBK	WrL	1991	A9 Eynsham
H424FUD	Volvo FL6.14	Excalibur CBK	WrL	1991	B9 Watlington

New to Oxfordshire in 1995, this Volvo FL6 with Excalibur CBK Rescue Tender bodywork is based at Kidlington fire station. An impressive array of specialist equipment is carried by M656TWL, including air, hydraulic and electrical cutting, lifting, jacking and spreading equipment. Beneath the tarpaulin on the appliance roof is an inflatable boat. *Simon Pearce*

J813MFC	Volvo FL6.14	Excalibur CBK	WrL	1991	B5 Goring
J814MFC	Volvo FL6.14	Excalibur CBK	WrL	1991	A11 Burford
J815MFC	Volvo FL6.14	Excalibur CBK	WrL	1991	A2 Hook Norton
J816MFC	Volvo FL6.14	Excalibur CBK	WrL	1991	A4 Charlbury
Pod	Frazer	Damage Control Unit	DCU	1992	B1 Rewley Road, Oxford
Pod	Frazer	Incident Support UNit (Cav)	ISU	1992	A6 Kidlington
K195UJO	Volvo FL6.14	Excalibur CBK	WrL	1992	B6 Henley on Thames
K196UJO	Volvo FL6.14	Excalibur CBK	WrL	1992	B6 Henley on Thames
K197UJO	Volvo FL6.14	Excalibur CBK	WrL	1992	A12 Bampton
K198UJO	Volvo FL6.14	Excalibur CBK	WrL	1992	B4 Wantage
K199UJO	Volvo FL6.14	Excalibur CBK	WrL	1992	B8 Wheatley
Pod	Frazer	Chemical Incident Unit	CIU	1993	B1 Rewley Road, Oxford
Pod	Frazer	Publicity Unit	PubU	1993	A6 Kidlington
L337EFC	Volvo FL6.14	Excalibur CBK	WrL	1993	Training Centre
L338EFC	Volvo FL6.14	Excalibur CBK	WrL	1993	B11 Wallingford
L339EFC	Volvo FL6.14	Excalibur CBK	WrL	1993	A3 Chipping Norton
L340EFC	Volvo FL6.14	Excalibur CBK	WrL	1993	A3 Chipping Norton
Pod	Frazer	General Purpose Van	GPV	1994	A6 Kidlington
M656TWL	Volvo FL6.14	Excalibur CBK	RT	1995	A6 Kidlington
M657TWL	Volvo FL6.14	Excalibur CBK	WrL	1994	A6 Kidlington
M658TWL	Volvo FL6.14	Excalibur CBK	WrL	1994	B2 Abingdon
M659TWL	Volvo FL6.14	Excalibur CBK	WrL	1994	B2 Abingdon
M660TWL	Volvo FL6.14	Excalibur CBK	WrL	1994	A10 Witney
M661TWL	Volvo FL6.14	Excalibur CBK	WrL	1994	A10 Witney
Pod	Frazer	Flat Bed Lorry	FBL	1995	A6 Kidlington
N436CBW	Volvo FL6.14	Excalibur CBK	WrL	1995	A1 Banbury
N437CBW	Volvo FL6.14	Excalibur CBK	WrL	1995	A1 Banbury
N438CBW	Volvo FL6.14	Excalibur CBK	WrL	1995	B1 Rewley Road, Oxford
N439CBW	Volvo FL6.14	Excalibur CBK	WrL	1995	B1 Rewley Road, Oxford
P281UJO	Volvo FL6.14	Excalibur CBK	WrL	1996	B1 Rewley Road, Oxford
P282UJO	Volvo FL6.14	Excalibur CBK	WrL	1996	B12 Didcot
P283UJO	Volvo FL6.14	Excalibur CBK	WrL	1996	B12 Didcot
P284UJO	Volvo FL6.14	Excalibur CBK	WrL	1996	B10 The Slade, Oxford
P285UJO	Volvo FL6.14	Excalibur CBK	WrL	1996	B10 The Slade, Oxford

A four year old Volvo FL6 truck chassis was acquired by Oxfordshire Fire Service in 1992. A refurbished water tank by Eagle Engineering (originally on a 1978 Dodge WrC with Bedfordshire) was mounted upon this Volvo, with additional bodywork by Frazer and Oxfordshire FS workshops. The completed appliance now serves as the brigade's Water Carrier. *Simon Pearce*

SHROPSHIRE FIRE & RESCUE SERVICE

Shropshire Fire & Rescue Service, St Michael's Street, Shrewsbury SY1 2HJ

1	K731AUJ	Land Rover Defender 110 Tdi	Saxon Sanbec	L4P	1992	Tweedale
2	K720BUJ	Volvo FL6-14	McDonald Kane Ground Loader	PM	1992	Telford Central
3	K414ENT	Dennis Rapier TF202	John Dennis	WrL	1993	Shrewsbury
4	K415ENT	Dennis Rapier TF202	John Dennis	RPL	1993	Ludlow
5	K416ENT	Dennis Rapier.TF202	John Dennis	RPL	1993	Bridgnorth
6	K417ENT	Dennis Rapier TF202	John Dennis	WrL	1993	Telford Central
8	L106JNT	Volvo FL6-18	Multilift	PM	1994	Shrewsbury
9	L107JNT	Volvo FL6-18	Multilift	PM	1994	Shrewsbury
10	Pod	Multilift/Genetech/Holland	Heavy Pumping Unit	HPU	1994	Shrewsbury

Most of Shropshires Prime Movers are based upon the Volvo FL6 chassis type.
Fleetnumber 8 is an FL6.18 model and was photographed with the Heavy Pumping Unit pod (Fleetnumber 10) which, additionally, carries 1 kilometre of hose. The pod was built by Genetech and Holland Fire System, 8 and 10 are both stationed at Shrewsbury.
Clive Shearman

A second-hand appliance with Shropshire Fire Service is their Turntable Ladder. New in 1987 to South Glamorgan it was sold to Shropshire after only two years. Now based at Shrewsbury, it has a MAN 12.192 chassis with Angloco bodywork. Unusually the Metz ladders have a maximum reach of around 27 metres.
Bill Potter

11	N344VUX	Volvo FL6.14	Saxon Sanbec/HIAB	RT	1996	Wellington
12	L668LUJ	Dennis Rapier TF203	John Dennis	RPL	1994	Whitchurch
13	L669LUJ	Dennis Rapier TF203	John Dennis	RPL	1994	Market Drayton
14	L670LUJ	Dennis Rapier TF203	John Dennis	RPL	1994	Oswestry
15	P685BUX	Land Rover Defender 110 Tdi	Land Rover	ICU	1997	Shrewsbury
16	N346VUX	Volvo FL6.14	Saxon Sanbec	WrL	1996	Wem
19	N347VUX	Volvo FL6.14	Saxon Sanbec	WrL	1996	Church Stretton
20	M683RAW	Dennis Rapier TF203	John Dennis	RPL	1995	Shrewsbury
21	M685RAW	Dennis Rapier TF203	John Dennis	RPL	1994	Telford Central
22	M684RAW	Dennis Rapier TF203	John Dennis	RPL	1994	Wellington
23	N348VUX	Volvo FL6.14	Saxon Sanbec	WrL	1996	Minsterley
27	N349VUX	Volvo FL6.14	Saxon Sanbec	WrL	1996	Much Wenlock
29	TNT755Y	Dodge G13C	Saxon Sanbec	WrL	1983	Fire Safety
30	TNT756Y	Dodge G13C	Saxon Sanbec	WrL	1983	Shrewsbury (R)
31	A85XUX	Dodge G13C	Saxon Sanbec	WrL	1983	Craven Arms
34	A456DNT	Dodge G13C	Saxon Sanbec	WrL	1984	Telford Central (R)
35	B481GUJ	Dodge G13C	Saxon Sanbec	WrL	1985	Shrewsbury (R)
36	B480GUJ	Dodge G13C	Saxon Sanbec	WrL	1985	Telford Central (R)
37	D337RUX	Renault-Dodge G13	Saxon Sanbec	WrL	1987	Hodnet
38	D338RUX	Renault-Dodge G13	Saxon Sanbec	WrL	1987	Newport
39	D339RUX	Renault-Dodge G13	Saxon Sanbec	WrL	1987	Wellington
70	F271DNT	Mercedes-Benz 917AF	Saxon Sanbec	ET	1988	Wellington (R)
71	F272DNT	Renault-Dodge G13	Saxon Sanbec	WrL	1988	Ellesmere
72	F273DNT	Renault-Dodge G13	Saxon Sanbec	WrL	1988	Baschurch
73	F274DNT	Renault-Dodge G13	Saxon Sanbec	WrL	1988	Clun
74	F275DNT	Renault-Dodge G13	Saxon Sanbec	WrL	1988	Prees
75	F761FAW	GMC 3500 4x4	Telehoist/Shropshire FB	RT	1988	Shrewsbury
76	D900ODW	MAN 12.192	Angloco/Metz	TL	1987	Shrewsbury
77	F861FNT	Bedford Astra Van	Bedford	FIU	1989	Shrewsbury
78	G688KAW	Dennis SS237	Saxon Sanbec	WrL	1989	Albrighton
79	G689KAW	Dennis SS237	Saxon Sanbec	WrL	1989	Training Centre
80	G690KAW	Dennis SS237	Saxon Sanbec	WrL	1989	Cleobury Mortimer
81	G691KAW	Dennis SS237	Saxon Sanbec	WrL	1989	Bishop's Castle
83	G855LNT	Land Rover 110 Tdi	Excalibur CBK	L4P	1990	Bridgnorth
84	G817LUX	Land Rover 110 Tdi	Excalibur CBK	L4P	1990	Oswestry
85	H421PNT	Dennis SS237	Saxon Sanbec	WrL	1990	Shrewsbury
86	H422PNT	Dennis SS237	Saxon Sanbec	WrL	1990	Shrewsbury (R)
87	H423PNT	Dennis SS237	Saxon Sanbec	WrL	1990	Oswestry
88	H424PNT	Dennis SS237	Saxon Sanbec	WrL	1990	Bridgenorth
90	H751SNT	Dennis SS234	McDonald Kane Groundloader	PM	1991	Shrewsbury
91	H294SUJ	Land Rover Defender 110 Tdi	Excaliber CBK	L4P	1991	Market Drayton
92	H295SUJ	Land Rover Defender 110 Tdi	Excaliber CBK	L4P	1991	Church Stretton
93	H296SUJ	Land Rover Defender 110 Tdi	Land Rover	PCV	1991	Fire Safety
94	H411SUX	Dennis F127	Saxon/Simon Snorkel ST240-S	AP	1991	Workshops
95	Trailer		Fire Safety	ExU	1991	Tweedale
96	Pod	McDonald Kane	Incident Support Unit	ISU	1991	Telford Central
97	Pod	McDonald Kane	Emergency Feeding Unit	CaV	1991	Telford Central
99	J661WNT	Dennis SS237	John Dennis	RPL	1992	Newport
100	J662WNT	Dennis SS237	John Dennis	WrL	1992	Market Drayton
101	J663WNT	Dennis SS237	John Dennis	WrL	1992	Ludlow
102	J561XAW	Volvo FL6-14	McDonald Kane Groundloader	PM	1992	Telford Central
103	J57XNT	Dennis SS237	Excaliber CBK	WrL	1992	Tweedale
104	J58XNT	Dennis SS237	Excaliber CBK	WrL	1992	Whitchurch
105	Pod	McDonald Kane/Shropshire	Technical Support/Control Unit	TSU	1991	Wellington
106	Boat		Boat and Trailer	WRU	19..	Shrewsbury
	Pod	W.H.Bence/Multilift	B A Training	BA(T)	1994	Shrewsbury
	Pod	W.H.Bence/Multilift	B A Training	BA(T)	1994	Shrewsbury

Notes: 75 was new to South Glamorgan County Fire Service, it was acquired in 1989.

SOMERSET FIRE BRIGADE

Somerset Fire Brigade, Hestercombe House, Cheddon Fitzpaine, Taunton TA2 8LQ

SYC124L	Range Rover 6x4	Carmichael	L4P	1973	Museum, Taunton
SYA885R	Land Rover Series III 88	Land Rover	L4V	1977	03 Dulverton
SYA886R	Land Rover Series III 88	Land Rover	L4V	1977	10 Nether Stowey
SYA887R	Land Rover Series III 88	Land Rover	L4V	1977	29 Castle Cary
VYB72W	Bedford TKG	HCB-Angus HSC	WrL	1981	Reserve
VYB73W	Bedford TKG	Saxon Sanbec(1991)	WrT	1981	Driver Trainer
YYA830X	Bedford TKG	HCB-Angus HSC	WrT	1981	28 Shepton Mallet
YYA831X	Bedford TKG	HCB-Angus HSC	WrL	1981	Reserve
YYA832X	Bedford TKG	HCB-Angus HSC	WrT	1981	Reserve
YYA833X	Bedford TKG	Wincanton	HFST	1981	02 Bridgwater
YYA836X	Land Rover Series III 88	Land Rover	L4V	1981	30 Wincanton
YYA837X	Land Rover Series III 88	Land Rover	L4V	1981	09 Williton
YYA838X	Bedford TKG	Wreckers International	RecV	1981	Headquarters

Above: **Pumping appliance purchases for Somerset standardised on Dodge in the 1970s, Bedford (1980-84) and then reverted to Dodge/Renault-Dodge until that marque was withdrawn from the market. Since 1989, full sized appliances have had Mercedes-Benz chassis, the latest being four 1124F models with bodywork by Saxon Sanbec. Bridgewater received N943BYD, shown here.**
Robert Smith

Opposite, top: **One of the later-style od bodywork associated with Bedford appliances is the Saxon Sanbec-design shown here as applied to FYC92Y.** *Karl Sillitoe*

Opposite, bottom: **In an attempt to make the appliance more visible Somerset's G486LYC has received yellow frontal treatment. The Mercedes-Benz 711D carries Saxon Sanbec bodywork and is based at Yeovil.** *Simon Pearce*

YYA840X	Bedford TKG	Bovey Bodies	GPV	1981	Headquarters
FYC91Y	Bedford TKG	Saxon Sanbec	WrL	1982	10 Nether Stowey
FYC92Y	Bedford TKG	Saxon Sanbec	WrT	1982	Reserve
FYC93Y	Bedford TKG	Saxon Sanbec	WrL	1982	Training Centre
A618OYA	Bedford TKG	Saxon Sanbec	WrT	1983	33 Chard
A619OYA	Bedford TKG	Saxon Sanbec	WrL	1983	33 Chard
A620OYA	Bedford TKG	Saxon Sanbec	WrL	1983	21 Yeovil
B871XYA	Bedford TKG	Saxon Sanbec	WrL	1984	22 Crewkerne
B872XYA	Bedford TKG	Saxon Sanbec	WrL	1984	04 Burnham-on-Sea
B873XYA	Bedford TKG	Saxon Sanbec	WrT	1984	08 Wiveliscombe
C998DYB	Dodge G13C	Saxon Sanbec	WrL	1985	01 Taunton
C999DYB	Dodge G13C	Saxon Sanbec	WrT	1985	02 Bridgwater
C738HYC	Freight-Rover Sherpa 350 V8	Saxon Sanbec	FSU	1986	01 Taunton
C739HYC	Freight-Rover Sherpa 350 V8	Saxon Sanbec	FSU	1986	21 Yeovil
D454KYB	Ford Cargo 0813	Ford/Somerset FB	IU	1986	Headquarters
D520KYD	Dodge G13C	Saxon Sanbec	WrL	1986	06 Porlock
D521KYD	Dodge G13C	Saxon Sanbec	WrT	1986	11 Minehead
E432SYD	Renault-Dodge G13C	Saxon Sanbec	WrL	1987	24 Somerton
E433SYD	Renault-Dodge G13C	Saxon Sanbec	WrT	1987	25 Street

Two Steyr-Daimler-Puch Pinzgauer L6P appliances are used on the run at Somerset's Cheddar and Dulverton fire stations, and both appliances have bodywork by Saxon Sanbec. These machines have built a reputation as exceptional off-road performers, and are undoubtedly a useful asset to firefighters with difficult terrain to traverse.
Karl Sillitoe

Somerset's latest hydraulic platform is N826BYC, based at Bridgewater. A Mercedes-Benz 1824 chassis supports Saxon Sanbec bodywork and the latest version of the established SS263 booms by Simon, this variant having different types of jacks to the earlier models.
Robert Smith

E434SYD	Renault-Dodge G13C	Saxon Sanbec	WrL	1987	11 Minehead
E435SYD	Renault-Dodge G13C	Saxon Sanbec	WrL	1987	23 Martock
E436SYD	Land Rover 90 Tdi	Land Rover	L4V	1987	08 Wiveliscombe
E437SYD	Land Rover 90 Tdi	Land Rover	L4V	1987	11 Minehead
E438SYD	Land Rover 90 Tdi	Land Rover	L4V	1987	26 Glastonbury
E348YYD	Land Rover 90 Tdi	Land Rover	L4V	1987	06 Porlock
E349YYD	Land Rover 90 Tdi	Land Rover	L4V	1987	07 Wellington
E350YYD	Land Rover 90 Tdi	Land Rover	L4V	1987	23 Martock
F765BYD	Renault-Dodge G13C	Saxon Sanbec	WrT	1988	05 Cheddar
F766BYD	Renault-Dodge G13C	Saxon Sanbec	WrT	1988	23 Martock
F767BYD	Renault-Dodge G13C	Saxon Sanbec	WrT	1988	31 Frome
F768BYD	Renault-Dodge G13C	Saxon Sanbec	WrL	1988	31 Frome
F916HYA	Land Rover 90 Tdi	Land Rover	L4V	1988	05 Cheddar
F918HYA	Land Rover 90 Tdi	Land Rover	L4V	1988	31 Frome
F917HYA	Land Rover 90 Tdi	Land Rover	L4V	1989	27 Wells
G481LYC	Mercedes-Benz 1120F	Saxon Sanbec	WrT	1989	09 Williton
G482LYC	Mercedes-Benz 1120F	Saxon Sanbec	WrL	1989	09 Williton
G483LYC	Mercedes-Benz 1625/52	Saxon/Simon Snorkel ST240-S	HP	1989	01 Taunton
G484LYC	Mercedes-Benz 1625/52	Saxon/Simon Snorkel ST240-S	HP	1989	21 Yeovil
G485LYC	Mercedes-Benz 1622/45	Saxon Sanbec	WrC	1990	02 Bridgwater
G486LYC	Mercedes-Benz 711D	Saxon Sanbec	RT	1990	21 Yeovil
H357TYB	Land Rover 110 Tdi	Land Rover	L4V	1990	21 Yeovil
H358TYB	Land Rover 90 Tdi	Land Rover	L4V	1990	24 Somerton
H359TYB	Land Rover 110 Tdi	Land Rover	L4V	1990	02 Bridgwater
H361TYB	Mercedes-Benz 1120F	Saxon Sanbec	WrT	1990	28 Shepton Mallet
H362TYB	Mercedes-Benz 1120F	Saxon Sanbec	WrT	1990	21 Yeovil
H363TYB	Mercedes-Benz 1120F	Saxon Sanbec	WrL	1990	07 Wellington
H364TYB	Mercedes-Benz 1120F	Saxon Sanbec	WrL	1990	29 Castle Cary
J51BYB	Mercedes-Benz 1120F	Saxon Sanbec	WrL	1991	10 Nether Stowey
J52BYB	Mercedes-Benz 1120F	Saxon Sanbec	WrT	1991	01 Taunton
J53BYB	Mercedes-Benz 1120F	Saxon Sanbec	WrL	1991	03 Dulverton
J54BYB	Mercedes-Benz 1120F	Saxon Sanbec	WrT	1991	32 Ilminster
J56BYB	Land Rover Defender 90 Tdi	Land Rover	L4V	1991	25 Street
J57BYB	Land Rover Defender 90 Tdi	Land Rover	L4V	1991	33 Chard
J58BYB	Mercedes-Benz 1722/45	Saxon Sanbec	WrC	1991	21 Yeovil
J59BYB	Steyr Pinzgauer 718K	Saxon Sanbec	L6P	1991	03 Dulverton
J61BYB	Mercedes-Benz 814D	Saxon Sanbec	CU	1991	02 Bridgwater
J493EYA	Land Rover Defender 90 Tdi	Land Rover	L4V	1992	04 Burnham-on-Sea
J494EYA	Land Rover Defender 90 Tdi	Land Rover	L4V	1992	22 Crewkerne
J495EYA	Land Rover Defender 110 Tdi	Land Rover	L4V	1992	01 Taunton
K801GYD	Mercedes-Benz 1124F	Saxon Sanbec	WrL	1993	07 Wellington
K802GYD	Mercedes-Benz 1124F	Saxon Sanbec	WrT	1993	04 Burnham-on-Sea
K803GYD	Mercedes-Benz 1124F	Saxon Sanbec	WrT	1993	26 Glastonbury
K804GYD	Mercedes-Benz 1124F	Saxon Sanbec	WrT	1993	27 Wells
K805GYD	Mercedes-Benz 814L	Saxon Sanbec	RT	1993	01 Taunton
L898NYA	Mercedes-Benz 814L	Saxon Sanbec	C/WrT	1994	27 Wells
L899NYA	Steyr Pinzgauer 718K	Saxon Sanbec	L6P	1994	05 Cheddar
L901NYA	Mercedes-Benz 1124F	Saxon Sanbec	WrL	1994	02 Bridgwater
L902NYA	Mercedes-Benz 1124F	Saxon Sanbec	WrL	1994	21 Yeovil
N436AYD			GPL	1995	Headquarters
N943BYD	Mercedes-Benz 1124F	Saxon Sanbec	WrL	1995	02 Bridgwater
N944BYD	Mercedes-Benz 1124F	Saxon Sanbec	WrL	1995	21 Yeovil
N945BYD	Mercedes-Benz 1124F	Saxon Sanbec	WrL	1995	01 Taunton
N946BYD	Mercedes-Benz 1124F	Saxon Sanbec	WrL	1995	28 Shepton Mallet
N826BYC	Mercedes-Benz 1824	Saxon/Simon Snorkel SS263	HP	1996	02 Bridgwater
N827BYC	?	?	PCV	1996	Headquarters
N828BYC	Land Rover Defender 110 Tdi	Land Rover	LUV	1996	11 Minehead
N829BYC	Land Rover Defender 110 Tdi	Land Rover	LUV	1996	21 Yeovil
N904CYC	Land Rover Defender 110 Tdi	Land Rover	LUV	1996	28 Shepton Mallet
N905CYC	Land Rover Defender 110 Tdi	Land Rover	LUV	1996	33 Chard
P95JYC	Mercedes-Benz 817	Saxon Sanbec	FSU	1997	
P96JYC	Mercedes-Benz 817	Saxon Sanbec	FSU	1997	
P	Mercedes-Benz 1124F	Saxon Sanbec	WrL	1997	On order
P	Mercedes-Benz 1124F	Saxon Sanbec	WrL	1997	On order
P	Mercedes-Benz 1124F	Saxon Sanbec	WrL	1997	On order
P	Mercedes-Benz 1124F	Saxon Sanbec	WrL	1997	On order

Note: D454KYB was previously with the SCC Environment Dept.

SOUTH WALES FIRE SERVICE

Lanelay Hall, Talbot Green, Pontyclun, Glamorgan CF7 9XA

(G - acquired from Gwent FB; M - acquired from Mid Glamorgan FS; S - acquired from South Glamorgan CFS)

M	2366	Merryweather GEM	Horse Drawn Steamer	P	1902	Preserved
M	HCJ839	Dennis F12	Dennis	PE	1956	Preserved
M	JNY537D	Diamond T	Recovery truck	RecV	19..	Preserved
G	KDW597P	Land Rover Series III 109	Land Rover/Gwent FB	L4P	1975	39 Blaenavon
G	KDW598P	Land Rover Series III 109	Land Rover/Gwent FB	L4P	1975	33 New Inn, Pontypool
G	OBO941R	Land Rover Series III 109	Land Rover/Gwent FB	L4P	1976	36 Blaina
M	TAX782S	ERF 84PF	ERF Fire Fighter	WrT	1977	Fire Prevention
M	Pod	Ray Smith/Powell Duffryn	Canteen Unit	CaV	1977	14 Pontyclun
M	Pod	Powell Duffryn	Chemical Incident Unit	CIU	1977	10 Pontypridd
M	Pod	Powell Duffryn	Salvage Unit	ST	1977	01 Bridgend
M	Pod 1	Powell Duffryn	Drop Side Lorry	GPV	1977	17 Aberdare
M	Pod 2	Powell Duffryn	Water Carrier	WrC	1977	01 Bridgend
M	Pod 3	Powell Duffryn	Oil Pollution Pod	OPP	1978	02 Porthcawl
M	Pod 4	Powell Duffryn	Flat Bed Lorry	FBL	1978	01 Bridgend
S	AUH289T	Dodge K850	HCB-Angus	FoT	1979	Withdrawn
S	BDW410T	Dodge K850	HCB-Angus	FoT	1979	Withdrawn
G	JBO14W	Land Rover Series III 109	HCB-Angus	L4P	1980	27 Tredegar
G	LNY617X	Shelvoke & Drewry WY	Angloco/Simon Snorkel SS263	HP	1980	32 Cwmbran
G	MUH545X	Ford Transit 190	Merry Weather/Gwent FB (1993)	CaV	1981	40 Abergavenny
M	Pod	Powell Duffryn/Hamworthy (1994)	BA Unit	BAU	1981	01 Bridgend
M	Pod	Powell Duffryn	Lighting & Test Unit	LTU	1982	17 Aberdare
M	MTH127X	Dodge G1613	Angloco/Simon Snorkel SS85	HP	1982	21 Merthyr Tydfil
M	NDW220X	Dodge G1313	Cheshire FE/Mid Glamorgan FB	FoT	1982	24 Caerphilly

Local government reorganisation took place throughout Wales on 1st April 1996 and this resulted in the formation of three new Local Fire Authorities from the previous eight county brigades. Gwent, Mid Glamorgan and South Glamorgan brigades now form the South Wales Fire Service. The last new special appliance for South Glamorgan was M830JAX at Penarth. This Volvo FL7/Bedwas Operational Support Unit has a Moffett Mounty All Terrain forklift truck. *Richard Eversden*

M	NDW221X	Dodge G1313	Cheshire FE/Mid Glamorgan FB	FoT	1982	01 Bridgend
S	OBO99X	Land Rover Series III 109	Cheshire Fire Engineering	L4P	1982	Withdrawn
S	OBO100X	Land Rover Series III 109	Cheshire Fire Engineering	L4P	1982	Ely, Cardiff
S	OBO101X	Land Rover Series III 109	Cheshire Fire Engineering	L4P	1982	Withdrawn
G	OWO332Y	Land Rover Series III 109	Saxon Sanbec	L4P	1982	44 Caldicot
G	PHB642Y	Land Rover Series III 109	Saxon Sanbec	L4P	1982	38 Brynmawr
G	PHB643Y	Land Rover Series III 109	Saxon Sanbec	L4P	1982	45 Malpas, Newport
G	PHB644Y	Land Rover Series III 109	Saxon Sanbec	L4P	1982	42 Usk
G	PHB645Y	Land Rover Series III 109	Saxon Sanbec	L4P	1982	29 Cefn Fforest, Blackwood
G	PHB646Y	Land Rover Series III 109	Saxon Sanbec	L4P	1982	35 Abertillery
G	PHB647Y	Land Rover Series III 109	Saxon Sanbec	L4P	1982	31 Risca
S	SKG271Y	Leyland Terrier	Powell Duffryn Rolonof	PM	1982	Whitchurch, Cardiff
S	TNY321Y	Land Rover Series III 109	Carmichael	L4P	1982	Withdrawn
M	UDW973Y	Bedford MK 4x4	Flat Bed Lorry	GPV	1982	Training School
S	TDW718Y	Dodge G13C	Mountain Range	WrL	1983	Training Centre
S	TDW719Y	Dodge G13C	Mountain Range	WrL	1983	Reserve
S	TDW720Y	Dodge G13C	Mountain Range	WrL	1983	Training Centre
M	Pod	Powell Duffryn	Accommodation Unit	AcU	1983	17 Aberdare
M	YAO457Y	DAF 3300	Wreckers International	RecV	1983	Workshops
S	B731AKG	Dodge S66C	Mountain Range	DCU	1984	Whitchurch, Cardiff
S	B457BHB	Dodge G16C	Carmichael/Magirus DL30	TL	1984	Cardiff Central
S	B458BHB	Dodge G13C	Mountain Range	WrL	1984	Reserve
S	B459BHB	Dodge G13C	Mountain Range	WrL	1984	Reserve
G	B806BKG	Land Rover 110	Saxon Sanbec	L4RP	1984	43 Chepstow
G	B807BKG	Land Rover 110	Saxon Sanbec	L4RP	1984	41 Monmouth
G	B600CBO	Dodge G13C	Saxon Sanbec	WrL	1984	??
G	B604CBO	Dodge G13C	Saxon Sanbec	WrL	1984	Reserve
G	B954CNY	Land Rover 110	Saxon Sanbec	L4P	1984	34 Abersychan
M	B481DMB	Dodge G13C	Saxon Sanbec	WrL	1985	Training School
M	B482DMB	Dodge G13C	Saxon Sanbec	WrL	1985	Reserve
M	B483DMB	Dodge G13C	Saxon Sanbec	WrL	1985	Reserve
S	C762FWD	Dodge G13C	Carmichael Fire Chief	WrL	1985	Cowbridge
G	C937GNY	Dodge G13C	Saxon Sanbec	WrL	1985	Reserve
G	C938GNY	Dodge G13C	Saxon Sanbec	WrL	1985	Reserve
G	C939GNY	Dodge G13C	Saxon Sanbec	WrL	1985	Reserve
G	C940GNY	Dodge G13C	Saxon Sanbec	WrL	1985	29 Cefn Fforest, Blackwood
G	C941GNY	Bedford TL570 Articulated	Reeve Burgess	TU	1985	45 Malpas, Newport
G	C551JBO	Bedford TL860	Saxon Sanbec	RT	1985	Reserve
G	C552JBO	Bedford TL860	Saxon Sanbec	RT	1985	Reserve
M	C992JLG	Dodge G13C	Saxon Sanbec	WrL	1985	Reserve
M	C993JLG	Dodge G13C	Saxon Sanbec	WrL	1985	Reserve
M	C860JMA	Dodge G13C	Saxon Sanbec	WrL	1985	08 Maesteg
M	C861JMA	Dodge G13C	Saxon Sanbec	WrL	1985	Reserve
M	C862JMA	Dodge G13C	Saxon Sanbec	WrL	1985	Reserve
G	Trailer	Lynton Commercials	Fire Safety Semi-Trailer	FSU	1985	45 Malpas, Newport
S	C125ETG	Land Rover 110	Hoskins	SRU	1986	Barry
G	D665MBO	Dodge G13C	Saxon Sanbec	WrL	1986	Reserve
G	D666MBO	Dodge G13C	Saxon Sanbec	WrL	1986	Driving School
G	D667MBO	Dodge G13C	Saxon Sanbec	WrL	1986	43 Chepstow
G	D668MBO	Dodge G13C	Saxon Sanbec	WrL	1986	Reserve
S	D923MDW	Land Rover 110	Saxon Sanbec	L4P	1986	Whitchurch, Cardiff
S	D876MHB	Dodge G13C	Saxon Sanbec	WrL	1986	Llantwit Major
S	D937ONY	Dodge G60	Holmes	HL	1986	Llantwit Major
M	Pod	Ray Smith/Powell Duffryn	Control Unit	CU	1986	14 Pontyclun
M	D128ERW	Volvo FL6.16	Powell Duffryn	PM	1987	10 Tonypandy
M	D129ERW	Volvo FL6.16	Powell Duffryn	PM	1987	17 Aberdare
M	D368JNX	Dennis F127	Saxon/Simon Snorkel SS263	HP	1987	01 Bridgend
S	D527MHB	Renault-Dodge G13C	HCB-Angus	WrL	1987	Penarth
M	D419TCA	Renault-Dodge G13C	Saxon Sanbec	WrL	1987	24 Caerphilly
M	D420TCA	Renault-Dodge G13C	Saxon Sanbec	WrL	1987	Training School
G	E411RBO	Renault-Dodge G13C	Saxon Sanbec	WrL	1987	39 Blaenavon
G	E412RBO	Renault-Dodge G13C	Saxon Sanbec	WrL	1987	41 Monmouth
G	E413RBO	Renault-Dodge G13C	Saxon Sanbec	WrL	1987	34 Abersychan
M	E45RKG	Ford Transit 160/County 4x4	Hayes & Ward	L4V	1987	24 Caerphilly
M	E46RKG	Ford Transit 160/County 4x4	Hayes & Ward	L4V	1987	01 Bridgend
M	E47RKG	Ford Transit 160/County 4x4	Hayes & Ward	L4V	1987	Reserve
M	E48RKG	Ford Transit 160/County 4x4	Hayes & Ward	L4V	1987	21 Merthyr Tydfil
M	E49RKG	Ford Transit 160/County 4x4	Hayes & Ward	L4V	1987	Training School
M	E177SUH	Volvo FL6.17	Brimec Slide-back Loader	PM	1987	Workshops
S	E40TUH	Land Rover 110	Saxon Sanbec	L4P	1987	Cowbridge
S	E721UTX	Renault-Dodge G13C	HCB-Angus	WrL	1987	Training Centre
S	F474BHB	Volvo FL6.14	Fulton & Wylie	WrL	1988	Whitchurch, Cardiff

G	F652BUH	Renault-Dodge G16	Saxon Sanbec	FoT	1988	47 Duffryn, Newport
M	F202XNY	Volvo FL6.14	Saxon Sanbec	WrL	1988	02 Porthcawl
M	F203XNY	Volvo FL6.14	Saxon Sanbec	WrL	1988	05 Pencoed
M	F204XNY	Volvo FL6.14	Saxon Sanbec	WrL	1988	03 Kenfig Hill
M	F985XTG	Iveco Ford Cargo 1415	Powell Duffryn	PM	1988	01 Bridgend
M	F629AUH	Volvo FL6.14	Saxon Sanbec	WrL	1989	10 Tonypandy
M	F630AUH	Volvo FL6.14	Saxon Sanbec	WrL	1989	08 Maesteg
M	F632AUH	Volvo FL6.14	Saxon Sanbec	WrL	1989	17 Aberdare
S	G446GHB	MAN G90 8-160	Fulton & Wylie	ET	1989	Ely, Cardiff
G	G320GNY	Scania G93M-250	Saxon Sanbec	WrL	1989	43 Chepstow
G	G321GNY	Scania G93M-250	Saxon Sanbec	WrL	1989	41 Monmouth
G	G999HBO	Scania G93M-250	Saxon Sanbec	WrL	1989	40 Abergavenny
S	G891KBO	Volvo FL6.14	Fulton & Wylie	WrL	1989	Cardiff Central
S	G892KBO	Volvo FL6.14	Fulton & Wylie	WrL	1989	Cardiff Central
M	G873JKG	Ford Transit 160/County 4x4	Ford	L4V	1990	Emergency Planning Department
M	G434KBO	Scania G93M 6x2	Saxon/Simon Snorkel ST240-S	HP	1990	15 Pontypridd
M	H273MHB	Ford Transit 160/County 4x4	Lonsdale	L4V	1990	08 Maesteg
M	H324MHB	Ford Transit 160/County 4x4	Lonsdale	L4V	1990	10 Tonypandy
M	H368MHB	Ford Transit 160/County 4x4	Lonsdale	L4V	1990	15 Pontypridd
S	H396MTX	Volvo FL6.14	Fulton & Wylie	WrL	1990	Barry
S	H397MTX	Volvo FL6.14	Fulton & Wylie	WrL	1990	Barry
G	H475NBO	Scania G93ML-250	Saxon Sanbec	WrL	1990	42 Usk
G	H476NBO	Scania G93ML-250	Saxon Sanbec	WrL	1990	32 Cwmbran
M	H251NUH	Ford Transit 160/County 4x4	Lonsdale	L4V	1990	23 Bargoed
G	H130NWO	Scania P93M-210	Alloy Transport Bodies	GPV	1990	30 Abercarn
G	H134NWO	Scania P93HL-250	Saxon Sanbec	WrL	1990	31 Risca
S	H126OKG	Volvo FL10	Saxon/Simon Snorkel ST300-S	AP	1990	Cardiff Central
S	H127OKG	Volvo FL6.14	HCB-Angus	WrL	1990	Ely, Cardiff
S	H135OKG	Volvo FL6.14	Powell Duffryn Rolonof	PM	1990	Cardiff Central
M	H299OTX	Volvo FL6.14	Saxon Sanbec	WrL	1991	01 Bridgend
M	H478OWO	Mercedes-Benz 1120AF	Saxon Sanbec	RT	1991	21 Merthyr Tydfil
M	H479OWO	Scania G93M-250	Saxon Sanbec	WrL	1991	21 Merthyr Tydfil
M	H481OWO	Mercedes-Benz 1120AF	HCB-Angus	RT	1991	01 Bridgend
M	H482OWO	Scania G93M-250	Saxon Sanbec	WrL	1991	01 Bridgend
M	H483OWO	Mercedes-Benz 1120AF	HCB-Angus	RT	1991	15 Pontypridd
M	H484OWO	Scania G93M-250	Saxon Sanbec	WrL	1991	15 Pontypridd
M	H485OWO	Mercedes-Benz 1120AF	Saxon Sanbec	WrL	1991	10 Tonypandy
M	H486OWO	Mercedes-Benz 1120AF	Saxon Sanbec	WrL	1991	13 Gilfach Goch
M	H487OWO	Mercedes-Benz 1120AF	Saxon Sanbec	WrL	1991	18 Abercynon
G	H591PTG	Scania G93ML-250	Saxon Sanbec	WrL	1991	27 Tredegar
G	H592PTG	Scania G93ML-250	Saxon Sanbec	WrL	1991	36 Blaing
M	J98SNY	Scania G93M-250	HCB-Angus	WrL	1991	14 Pontyclun
M	J652STX	Scania G93M-250	HCB-Angus	WrL	1991	24 Caerphilly
G	J564UDW	Mercedes-Benz 1120AF	Saxon Sanbec	WrL	1991	40 Abergavenny
S	J573UDW	MAN G90 8-160	Fulton & Wylie	ET	1992	Whitchurch, Cardiff
G	J734VHB	Scania P93HL-250	Saxon Sanbec	WrL	1992	38 Brynmawr
G	J735VHB	Scania P93HL-250	Saxon Sanbec	WrL	1992	Abercarn
G	J736VHB	Scania P93HL-250	Saxon Sanbec	WrL	1992	35 Abertillery
M	J250WBO	Mercedes-Benz 1124AF	HCB-Angus	WrL	1992	06 Ogmore Vale
M	J251WBO	Mercedes-Benz 1124AF	HCB-Angus	WrL	1992	22 Rhymney
M	J252WBO	Mercedes-Benz 1124AF	HCB-Angus	WrL	1992	16 Hirwaun
M	J253WBO	Mercedes-Benz 1124AF	HCB-Angus	WrL	1992	21 Merthyr Tydfil
M	J254WBO	Mercedes-Benz 1124AF	HCB-Angus	WrL	1992	15 Pontypridd
M	Trailer	Keillor	Display Unit Semi-Trailer	DisU	1992	Fire Prevention Department
S	K691XWO	Volvo FS7.18	Saxon Sanbec	WrL	1992	Roath, Cardiff
M	K578YAX	Iveco Turbo Daily 49.12 Tractor	Keillor	TU	1992	Fire Prevention Department
G	K67YBO	Scania G93ML-250	Saxon Sanbec	WrL	1992	47 Duffryn, Newport
G	K68YBO	Scania G93ML-250	Saxon Sanbec	WrL	1992	44 Caldicot
G	K69YBO	Mercedes-Benz 814D	Optare StarRider/Saxon	CU	1992	32 Cwmbran
M	K475ATG	Renault D160	Wadham Stringer Vanguard	PCh	1993	17 Aberdare
M	K335BHB	ERF ES8.24RD2	Dairy Products	WrC	1993	15 Pontypridd
M	K934YNY	Mercedes-Benz 1124AF	Saxon Sanbec	WrL	1993	19 Treharris
G	L26DDW	Scania P93ML-250	Angloco	CIU	1993	32 Cwmbran
G	L27DDW	Scania P93ML-250	Saxon Sanbec	WrL	1993	45 Malpas, Newport
G	L29DDW	Scania P93M	Saxon/Simon Snorkel SS263	HP	1993	46 Maindee, Newport
M	L889DTG	Mercedes-Benz 1124AF	Saxon Sanbec	WrL	1993	07 Pontycymmer
M	L969FTX	Mercedes-Benz 1124AF	Saxon Sanbec	WrL	1994	23 Bargoed
M	L970FTX	Mercedes-Benz 1124AF	Saxon Sanbec	WrL	1994	09 Treorchy
M	M764JTG	Mercedes-Benz 1124AF	Saxon Sanbec	WrL	1994	08 Maesteg
M	M765JTG	Mercedes-Benz 1124AF	Saxon Sanbec	WrL	1994	17 Aberdare
M	M558JTX	Mercedes-Benz 1124AF	Saxon Sanbec	WrL	1994	19 Treharris

M659KDW from Whitchurch fire station in Cardiff, is seen at the University Hospital of Wales, having answered an emergency call . This WrL appliance was built by Saxon Sanbec onto a Volvo FS7 chassis. An uprated variant of the familiar FL6 type, the FS7 features a more powerful 7-litre engine, uprated brakes, and an increased gross vehicle weight. To accommodate the larger engine, the FS7s cab is mounted higher on the chassis. *John Jones*

S	M830JAX	Volvo FL7	Bedwas	OSU	1995	Penarth
S		Moffett Mounty All Terrain	*Normally carried by M830JAX*	FLT	1995	Penarth
G	M972JWO	Scania G93ML-250	Saxon Sanbec	WrL	1995	46 Maindee, Newport
G	M973JWO	Scania P93HKZ-250 4x4	Saxon Sanbec	RT	1995	37 Ebbw Vale
G	M974JWO	Scania P93HKZ-250 4x4	Saxon Sanbec	RT	1995	45 Malpas, Newport
M	M660KDW	Volvo FL6.14	Lacre PDE	PM	1995	14 Pontyclum
S	M659KDW	Volvo FS7.18	Saxon Sanbec	WrL	1995	Whitchurch
M	M96KWO	Ford Transit 90/County 4x4	Ford	L4V	1995	17 Aberdare
M	M132KWO	Iveco-Ford Daily Cargo 1405185	Flat Bed Lorry	FBL	1995	Training School
M	M101KBO	ERF ES8.24 RD2	Massey Tankers	WrC	1995	01 Bridgend
M	M841MBX	Land Rover 110 2.5D	Land Rover	L4V	1995	Workshops
G	N310NTG	Scania P93M-250	Angloco	WrL	1996	32 Cwmbran
G	N311NTG	Scania P93M-250	Angloco	WrL	1996	37 Ebbw Vale
G	N312NTG	Scania P93M-250	Angloco	WrL	1996	29 Cefn Fforest
G	N313NTG	Scania P93M-250	Angloco	WrL	1996	33 New Inn
S	N405NHB	Volvo FS7.18	Saxon Sanbec	WrL	1996	Cardiff Central
S	N406NHB	Volvo FS7.18	Saxon Sanbec	WrL	1996	Cardiff Central
S	N407NHB	Mercedes-Benz		SRU	1996	Barry
S	Pod		BA Unit	BAU	19..	Cardiff Central
S	Pod		Canteen Van	CaV	19..	Cardiff Central
S	Pod		Damage Control Van	DCU	19..	Cardiff Central
S	Pod		General Purpose Unit	GPV	19..	Cardiff Central
S	Pod		General Purpose Unit	GPV	19..	Cardiff Central
S	Pod		Incident Command Unit	ICU	19..	Cardiff Central
M	Pod	Powell Duffryn	Hose Layer	HL	1987	17 Aberdare

Notes:
JNY537D was acquired and registered in 1966. MUH545X was formerly a Control Unit. NDW220/221X were formerly Water Tender Ladders. C762FWD was orignally a Carmichael demonstrator. D527MHB was orginally a HCB-Angus demonstrator. G434KBO was orignally built to 4x2 layout

SOUTH YORKSHIRE FIRE & RESCUE SERVICE

South Yorkshire County Fire Brigade, Wellington Street, Sheffield S1 3FG

VWG511S	Dennis Delta II	Dennis/Simon Snorkel SS263	HP	1978	1 Barnsley
KKY940W	Dennis F127	Carmichael/Magirus DL30U	TL	1981	6 Doncaster
Pod	Wilson Masterbodies	Smoke Training Unit	BA(T)	1981	Training Centre
Pod	Williams of Cardiff	Decontamination Unit	DecU	1981	26 Tankersley
Pod	Williams of Cardiff	Salvage Tender	ST	1981	26 Tankersley
Pod	Williams of Cardiff	General Purpose Vehicle	GPV	1981	26 Tankersley
Pod	Angloco	1000 Gallon Tanker	WrC	1982	20 Brampton
PHL331X	Dennis F125	Carmichael/Magirus DL30U	TL	1982	18 Rotherham
PKU738X	Ford Transit	Angloco	RT	1982	Reserve
UKW586Y	Dodge G13	Wilson	DTV	1983	Driving School
VDT421Y	Dennis RS133	Dennis	WrL	1983	Training Centre
VDT422Y	Dennis RS133	Dennis	WrL	1983	Driving School
VDT424Y	Dennis RS133	Dennis	WrL	1983	21 Maltby
VDT426Y	Dennis RS133	Dennis	WrL	1983	Fire prevention department
A415YWA	Dennis RS133	Dennis	WrL	1983	Driving School
A417YWA	Dennis RS133	Dennis	WrL	1983	Training Centre
B47CKY	Ford Transit/County 4x4	Angloco	RT	1984	Training Centre
B48CKY	Dennis RS133	Dennis	ET	1984	1 Barnsley
B101DET	Dennis RS133	Dennis	RP	1984	Reserve
B102DET	Dennis RS133	Dennis	WrL	1984	Reserve
B103DET	Dennis RS133	Dennis	WrL	1984	Reserve
B104DET	Dennis RS133	Dennis	WrL	1984	Reserve
B105DET	Dennis RS133	Dennis	WrL	1984	Reserve
B106DET	Dennis RS133	Dennis	WrL	1984	Reserve

Barnsley Fire Station is home to a pair of Water Tender Ladders, an Emergency Tender and this Hydraulic Platform, VWG511S. This view illustrates the combination of Dennis Delta II chassis, Dennis bodywork and Simon Snorkel SS263 booms. *Graham Hopwood*

Pod	Transliner	Fire Prevention Exhibition Unit	FPU	1984	26 Tankersley
B642EET	Dennis F125	Carmichael/Magirus DLK23-12	TL	1985	13 Sheffield Central
B643EET	Dennis RS133	Dennis	ET	1985	13 Sheffield Central
B437FKW	Dennis RS133	Dennis	ET	1985	6 Doncaster
B536GWB	Dennis RS133	Dennis	WrL	1985	Reserve
B537GWB	Dennis RS133	Dennis	WrL	1985	??
B538GWB	Dennis F127	Saxon/Simon Snorkel SS263	HP	1985	13 Sheffield Central
C715HKW	Land Rover 110	Land Rover	L4V	1986	13 Sheffield Central
D302OWA	Dennis F127	Saxon/Simon Snorkel SS220	HP	1986	24 Elm Lane, Sheffield
D516PET	Dennis RS133	Angloco	WrL	1986	22 Kiveton Park
D517PET	Dennis RS133	Angloco	WrL	1986	27 Mosborough
D518PET	Dennis RS133	Angloco	WrL	1986	5 Penistone
D519PET	Dennis RS133	Angloco	WrL	1986	2 Royston
D482RKW	Dennis RS135	Carmichael	BAT	1987	28 Edlington Lane, Doncaster
E918WET	Dennis RS133	Carmichael	WrL	1988	12 Rossington
E919WET	Dennis RS133	Carmichael	WrL	1988	22 Kiveton Park
E920WET	Dennis RS133	Carmichael	WrL	1988	20 Brampton Bierlow
E999WET	Volvo FL6.14	Excalibur CBK	WrL	1988	15 Low Edges Road, Sheffield
E75XWE	Volvo FL6.14	Excalibur CBK	WrL	1988	8 Mexborough
E76XWE	Volvo FL6.14	Excalibur CBK	WrL	1988	14 Rivelin Valley Road, Sheffield
E77XWE	Volvo FL6.14	Excalibur CBK	WrL	1988	13 Sheffield Central
E78XWE	Volvo FL6.14	Excalibur CBK	WrL	1988	17 Stocksbridge
E79XWE	Volvo FL6.14	Excalibur CBK	WrL	1988	24 Elm Lane, Sheffield
E80XWE	Volvo FL6.14	Excalibur CBK	WrL	1988	6 Doncaster
E81XWE	Volvo FL6.14	Excalibur CBK	WrL	1988	10 Askern
E82XWE	Volvo FL6.14	Excalibur CBK	WrL	1988	26 Tankersley
E83XWE	Volvo FL6.14	Excalibur CBK	WrL	1988	1 Barnsley
E84XWE	Volvo FL6.14	Excalibur CBK	WrL	1988	5 Penistone
E251XWE	Ford Transit	TC Harrison	GPV	1988	13 Sheffield Central
F999AWU	Volvo FL10	Flatbed	GPV	1989	Driving School
F854DWA	Mercedes-Benz 917AF	Carmichael	ET	1989	18 Rotherham
F616EWJ	ERF E6.18	Angloco	CU	1989	26 Tankersley
G542HKW	Volvo FL6.14	Excalibur CBK	WrL	1989	18 Rotherham
G617JHL	Volvo FL6.14	Excalibur CBK	WrL	1989	23 Darnall Road, Sheffield
G618JHL	Volvo FL6.14	Excalibur CBK	WrL	1989	16 Ringinglow Road, Sheffield
G619JHL	Volvo FL6.14	Excalibur CBK	WrL	1989	28 Edlington Lane, Doncaster
G620JHL	Volvo FL6.14	Excalibur CBK	WrL	1989	18 Rotherham
G629JHL	Volvo FL6.14	Excalibur CBK	WrL	1989	6 Doncaster
G630JHL	Volvo FL6.14	Excalibur CBK	WrL	1989	23 Darnall Road, Sheffield
H346SHE	Land Rover Defender 110	Land Rover	L4V	1990	West Division
H347SHE	Land Rover Defender 110	Land Rover	L4V	1990	East Division
H348SHE	Land Rover Defender 110	Land Rover	L4V	1990	East Division
H349SHE	Land Rover Defender 110	Land Rover	L4V	1990	East Division
J989DRH	Renault G300	Holmes	GPV	1991	Driving School
K402EET	GMC K31403 4x4	Angloco	RRRT	1993	Reserve
K406EET	Mercedes-Benz 1124F	Saxon Sanbec	RP	1993	7 Adwick-le-Street
K407EET	Mercedes-Benz 1124F	Saxon Sanbec	WrL	1993	13 Sheffield Central
K408EET	Mercedes-Benz 1124F	Saxon Sanbec	WrL	1993	25 Mansfield Road, Sheffield
K410EET	Mercedes-Benz 1124F	Saxon Sanbec	WrL	1993	9 Thorne
K411EET	Mercedes-Benz 1124F	Saxon Sanbec	WrL	1993	1 Barnsley
K412EET	Mercedes-Benz 1124F	Saxon Sanbec	WrL	1993	21 Maltby
K413EET	Mercedes-Benz 1124F	Saxon Sanbec	WrL	1993	2 Royston
K386FET	GMC K31403 4x4	Angloco	RRRT	1993	28 Edlington Lane, Doncaster
K838JWB	GMC K31403 4x4	Saxon Sanbec	RRRT	1993	25 Mansfield Road, Sheffield
K839JWB	GMC K31403 4x4	Saxon Sanbec	RRRT	1993	26 Tankersley
L345LET	Mercedes-Benz 2527	Swain & Bradshaw	OSU	1994	26 Tankersley
	Moffett Mounty M2403	*Normally carried by L345LET*	FLT	1994	26 Tankersley
L704MKY	Mercedes-Benz 1820	Lacre PDE	PM	1994	26 Tankersley
L705MKY	Mercedes-Benz 1820	Lacre PDE	PM	1994	26 Tankersley

For many years the cab design on new models in the Dennis range tended to bear a distinct resemblance to their predecessors. By the time that PHL331X was built, most of the Dennis range had changed over to using the Ogle-designed safety cab as used on all but the earliest RS type appliances. For those types that required a low-line cab, the earlier design continued to be used until the mid 1980s. Allocated to Rotherham Fire Station, this machine marries the low-line F125 chassis Carmichael bodywork and Magirus DL30U ladder sections. *Karl Sillitoe*

The last Dennis appliances to enter the South Yorkshire fleet arrived in 1988 and were a trio of Carmichael-bodied RS133 Water Tender Ladders. Subsequent deliveries have been on Volvo and Mercedes-Benz chassis. Here is E920WET which is used on the run at Brampton Bierlow. *Karl Sillitoe*

STAFFORDSHIRE FIRE & RESCUE SERVICE

Staffordshire Fire and Rescue Service, Service HQ, Pirehill, Stone, Staffs ST15 0BS

		Shand Mason	Double Vent Steam Pump	P	1875	Preserved, Burton on Trent
111	D998AVT	Dennis RS135	John Dennis	PRL	1986	Tutbury
113	M113WVT	Dennis SS239	John Dennis	WrT	1995	Newcastle under Lyme
114	M114WVT	Dennis SS239	John Dennis	PRL	1995	Longton
115	M115WVT	Dennis SS239	John Dennis	WrT	1995	Burton on Trent
119	C52TVT	Dennis RS135	John Dennis	WrT	1985	Reserve
120	CRE257Y	Dennis RS133	Dennis	WrT	1982	Wombourne
122	A334MBF	Dennis RS135	John Dennis	WrT	1984	Cheadle
123	B65PRF	Dennis RS135	John Dennis	PRL	1985	Longnor
129	B66PRF	Dennis RS135	John Dennis	PRL	1985	Gnosall
130	B67PRF	Dennis RS135	John Dennis	WrL	1985	Reserve
132	C51TVT	Dennis RS135	John Dennis	WrT	1985	Rugeley
133	C681WRE	Dennis RS135	John Dennis	WrL	1986	Reserve
137	C682WRE	Dennis RS135	John Dennis	PRL	1986	Biddulph
139	D22BEH	Dennis RS135	John Dennis	PRL	1986	Reserve
142	N142DVT	Dennis Sabre	John Dennis	RL	1995	Hanley
145	E827HBF	Dennis RS135	John Dennis	PRL	1987	Reserve
146	E828HBF	Dennis RS135	John Dennis	PRL	1987	Brewood
147	H766HFA	Dennis RS137	John Dennis	WrT	1990	Lichfield
148	H767HFA	Dennis RS137	John Dennis	WrT	1990	Driver Trainer
149	H768HFA	Dennis RS137	John Dennis	WrT	1990	Driver Trainer
150	H769HFA	Dennis RS137	John Dennis	WrT	1990	Uttoxeter

Staffordshire's hydraulic platform appliance is 503 (B369OFA). This vehicle has the low line Dennis DF127 chassis, Dennis bodywork and Snorkel SS220 booms from Simon. Originally stationed at Leek, a recent reallocation saw the vehicle moved to Longton where it was photographed. *Karl Sillitoe*

151	H112KRE	Dennis RS137	John Dennis	WrT	1990	Tamworth
152	H113KRE	Dennis RS137	John Dennis	PRL	1990	Barton
153	H114KRE	Dennis RS137	John Dennis	PRL	1990	Codsall
154	J52RFA	Dennis RS237	John Dennis	PRL	1991	Chase Terrace
155	J53RFA	Dennis RS237	John Dennis	WrT	1991	Newcastle under Lyme
156	J54RFA	Dennis RS237	John Dennis	PRL	1991	Central Training Unit
157	J56RFA	Dennis RS237	John Dennis	WrT	1991	Burton on Trent
158	J57RFA	Dennis RS137	John Dennis	WrT	1991	Stafford
159	J58RFA	Dennis RS137	John Dennis	PRL	1991	Burslem
160	J59RFA	Dennis RS137	John Dennis	PRL	1991	Kidsgrove
161	J61RFA	Dennis RS137	John Dennis	PRL	1991	Lichfield
162	J62RFA	Dennis RS137	John Dennis	WrT	1991	Cannock
163	D21BEH	Dennis RS135	John Dennis	PRL	1986	Ipstones
164	K768DFA	Dennis RS237	John Dennis	RL	1993	Burton on Trent
165	K769DFA	Dennis RS237	John Dennis	WrT	1993	Longton
166	K770DFA	Dennis RS237	John Dennis	PRL	1993	Tamworth
168	K771DFA	Dennis RS237	John Dennis	RL	1993	Leek
169	K772DFA	Dennis RS237	John Dennis	PRL	1993	Newcastle under Lyme
171	CRE255Y	Dennis RS133	Dennis	WrT	1982	Reserve
172	K773DFA	Dennis RS237	John Dennis	PRL	1993	Cannock
173	K774DFA	Dennis RS239	John Dennis	PRL	1993	Stafford
174	K775DFA	Dennis RS239	John Dennis	WrT	1993	Hanley
177	E830HBF	Dennis RS137	John Dennis	PRL	1987	Ashley
182	F817SFA	Dennis RS137	John Dennis	WrT	1988	Stone
183	F818SFA	Dennis RS137	John Dennis	WrT	1988	Abbots Bromley
184	F819SFA	Dennis RS137	John Dennis	WrT	1988	Burslem
185	F820SFA	Dennis RS137	John Dennis	WrT	1988	Central Training Unit
186	G21AVT	Dennis RS137	John Dennis	WrT	1989	Kidsgrove
187	G22AVT	Dennis RS137	John Dennis	WrT	1990	Leek
188	G23AVT	Dennis RS137	John Dennis	RL	1989	Wombourne
189	G24AVT	Dennis RS137	John Dennis	PRL	1989	Stone
190	G25AVT	Dennis RS137	John Dennis	PRL	1989	Rugeley
191	G26AVT	Dennis RS137	John Dennis	PRL	1989	Cheadle
192	G27AVT	Dennis RS137	John Dennis	PRL	1989	Uttoxeter
193	G28AVT	Dennis RS137	John Dennis	PRL	1989	Eccleshall
194	G29AVT	Dennis RS137	John Dennis	PRL	1989	Kinver
195	G30AVT	Dennis RS137	John Dennis	WrL	1989	Reserve
196	P	Dennis Sabre ML200	John Dennis	PRL	1997	
197	P	Dennis Sabre ML200	John Dennis	PRL	1997	
199	D999AVT	Dennis RS135	John Dennis	PRL	1986	Penkridge
261	K841YRE	Land Rover Defender 90 Tdi	Land Rover	L4V	1992	Leek
262	K842YRE	Land Rover Defender 90 Tdi	Land Rover	L4V	1992	Cannock
263	K843YRE	Land Rover Defender 90 Tdi	Land Rover	L4V	1992	Stafford
264	N264DBF	Land Rover Defender 90 Tdi	Land Rover	L4V	1995	Tamworth
265	N265DBF	Land Rover Defender 110 Tdi	Land Rover	CU	1995	Headquarters
300	JRE294	Leyland Hippo	Leyland	P	1939	Preserved, Uttoxeter
301	RFA868	Bedford TK	Merryweather	TL	1963	Preserved, Burton on Trent
500	A59HRE	Bedford TK	Flat Bed Lorry	GPV	1984	Driver Trainer
503	B369OFA	Dennis DF127	Dennis/Simon Snorkel SS220	HP	1984	Longton
505	JVT471V	Ford D1114	Benson	DLU	1979	Kidsgrove
506	A560LRE	Bedford TM 4x4	Carmichael	FoT/WrC	1984	Cannock
509	A561LRE	Bedford TM 4x4	Carmichael	FoT/WrC	1984	Leek
510	C930UVT	Bedford TL860	Fulton & Wylie	Cav	1985	Rugeley
512	B527URO	Jeep J20 6x4	Fulton & Wylie	RT	1984	Reserve
516	G317DRF	Dennis SS137	John Dennis	RT	1990	Newcastle under Lyme
517	G319DRF	Dennis SS137	John Dennis	RT	1990	Stafford
518	H168MEH	Crayford Argocat K18	Crayford Engineering	ATV	1991	Leek
519	H169MEH	Crayford Argocat K18	Crayford Engineering	ATV	1991	Cannock
520	J520PDH	AWD TL11-16	Lynton Commercials	ExU	1992	Fire safety
521	K531BEH	Crayford Argocat K18	Crayford Engineering	ATV	1992	Tamworth
522	L522NRF	MAN M17.272FL	MAN/RSG	PM	1993	Burslem
523	L523NRF	MAN M17.272FL	MAN/RSG	PM	1993	Cannock
524	Pod	Reynolds Boughton	Foam Tender	FoT	1993	Burslem
525	Pod	Reynolds Boughton	Foam Tender	FoT	1993	Cannock
534	PRE441W	Bedford TKG	Benson	DLU	1980	Burton on Trent
539	A363JRF	MAN 16-240	Angloco/Metz DL30	TL	1983	Burton on Trent
543	M543SRF	Scania P113H-320	Angloco/Bronto Skylift F30HDT	HP	1994	Stafford
547	E935JFA	Jeep J20 6x4	John Dennis	RT	1987	Lichfield

548	E784JEH	ERF Hi-Line E6.18	John Dennis	CU	1987	Leek
549	E783JEH	Dennis SS135	John Dennis	CIU	1987	Kidsgrove
550	F914SRE	Dennis SS137	John Dennis	SIU	1987	Burton on Trent
551	E282MRE	Dennis F127	John Dennis/Camiva EPA30	TL	1988	Hanley

Staffordshire Fire & Rescue Service operate five rescue tenders, three based on Jeep tri-axle vehicles while the other pair are based on Dennis SS137s. One of these with John Dennis bodywork is, 516 (G317DRF) which is based in Newcastle-under-Lyne and is shown here.
Bill Murray

In 1987 Staffordshire took delivery of this ERF Hi-Line E6-18 bodied as a Control Unit by John Dennis. Although ERF are attempting to re-enter the fire appliance market and have gained orders from several brigades, 548 (E784JEH) remains the only example with Staffordshire. It is based at Leek.
Malcolm Cook

STRATHCLYDE FIRE BRIGADE

Strathclyde Region Fire Brigade, Bothwell Road, Hamilton ML3 0EA

OGD142M	Land Rover Series II 109	Strathclyde FB	L4P	1972	F69 Craighouse V, Jura
CGE592S	Bedford TKD	Fulton & Wylie Fire Witch	M/WrT	1978	Headquarters
TDS71W	Bedford TKD	Fulton & Wylie Fire Witch	M/WrT	1981	North Reserve
DGD213X	Bedford KD	Fulton & Wylie Fire Witch	M/WrT	1981	F60 Scarinish, Tiree
EGD267X	Bedford KG	Cheshire Fire Engineering	WrL	1981	West Reserve
EGD270X	Bedford KG	Cheshire Fire Engineering	WrL	1981	West Reserve
EGD271X	Bedford KG	Cheshire Fire Engineering	WrL	1981	Headquarters
EGD272X	Bedford KG	Cheshire Fire Engineering	WrL	1981	North Reserve
EGD274X	Bedford KG	Cheshire Fire Engineering	WrL	1981	West Reserve
EGD275X	Bedford KG	Cheshire Fire Engineering	WrL	1981	D17 Muirkirk
EGD276X	Bedford KG	Cheshire Fire Engineering	WrL	1981	North Reserve
EGD277X	Bedford KG	Cheshire Fire Engineering	WrL	1981	F20 Campbeltown
GGB30X	Bedford KD	Fulton & Wylie Fire Witch	M/WrT	1982	F26 Garelochhead
LGD625Y	Bedford TK	Fulton & Wylie Fire Warrior	WrL	1982	East Reserve
LGD626Y	Bedford TK	Fulton & Wylie Fire Warrior	WrL	1982	North Reserve
LGD627Y	Bedford TK	Fulton & Wylie Fire Warrior	WrL	1982	Central Reserve
LGD628Y	Bedford TK	Fulton & Wylie Fire Warrior	WrL	1982	Central Reserve
LGD629Y	Bedford TK	Fulton & Wylie Fire Warrior	WrL	1982	Central Reserve
LGD630Y	Bedford TK	Fulton & Wylie Fire Warrior	WrL	1982	East Reserve
LGD631Y	Bedford TK	Fulton & Wylie Fire Warrior	WrL	1982	Central Reserve
LGD632Y	Bedford TK	Fulton & Wylie Fire Warrior	WrL	1982	Central Reserve
LGD633Y	Bedford TK	Fulton & Wylie Fire Warrior	WrL	1982	West Reserve
LGD634Y	Bedford TK	Fulton & Wylie Fire Warrior	WrL	1982	North Reserve
OYS145Y	Bedford KD	Fulton & Wylie Fire Witch	M/WrT	1983	F66 Port Ellen V, Islay
OYS146Y	Bedford KD	Fulton & Wylie Fire Witch	M/WrT	1983	F63 Bunessan V, Mull
OYS147Y	Bedford KD	Fulton & Wylie Fire Witch	M/WrT	1983	F23 Rothesay
OYS148Y	Bedford KD	Fulton & Wylie Fire Witch	M/WrT	1983	F15 Oban
TSJ132Y	Ford P100	Strathclyde FB	VSU	1983	F65 Iona V
TSJ133Y	Ford P100	Strathclyde FB	VSU	1983	D52 Blackwaterfoot V
TSJ134Y	Ford P100	Strathclyde FB	VSU	1983	D51 Lochranza V
A811XSJ	Bedford TK	Fulton & Wylie Fire Warrior	WrL	1983	Training Centre
A812XSJ	Bedford TK	Fulton & Wylie Fire Warrior	WrL	1983	Training Centre
A813XSJ	Bedford TK	Fulton & Wylie Fire Warrior	WrL	1983	D19 New Milns
A814XSJ	Bedford TK	Fulton & Wylie Fire Warrior	WrL	1983	Central Reserve
A815XSJ	Bedford TK	Fulton & Wylie Fire Warrior	WrL	1983	West Reserve
A816XSJ	Bedford TK	Fulton & Wylie Fire Warrior	WrL	1983	Central Reserve
A817XSJ	Bedford TK	Fulton & Wylie Fire Warrior	WrL	1983	D13 Colmonell
A818XSJ	Bedford TK	Fulton & Wylie Fire Warrior	WrL	1983	West Reserve
A819XSJ	Bedford TK	Fulton & Wylie Fire Warrior	WrL	1983	Central Reserve
A820XSJ	Bedford TK	Fulton & Wylie Fire Warrior	WrL	1983	West Reserve

Strathclyde Fire Brigade have several road rescue units which are equipped to attend road traffic accidents. A new generation of these vehicles are now being introduced, based on the Mercedes-Benz Sprinter model. N298NGG is based at Dumbarton.
Gavin Stewart

A31ASJ	Bedford TK	Fulton & Wylie Fire Warrior	WrL	1984	North Reserve
A32ASJ	Bedford TK	Fulton & Wylie Fire Warrior	WrL	1984	East Reserve
A33ASJ	Bedford TK	Fulton & Wylie Fire Warrior	WrL	1984	F21 Tignabruaich
A34ASJ	Bedford TK	Fulton & Wylie Fire Warrior	WrL	1984	East Reserve
A35ASJ	Bedford TK	Fulton & Wylie Fire Warrior	WrL	1984	North Reserve
A36ASJ	Bedford TK	Fulton & Wylie Fire Warrior	WrL	1984	D05 Dreghorn
A834WGG	Shelvoke & Drewry WY	Saxon/Simon Snorkel SS263	HP	1984	E01 Hamilton
B870AGD	Shelvoke & Drewry WY	Saxon/Simon Snorkel SS263	HP	1985	A04 North West Glasgow
B871AGD	Shelvoke & Drewry WY	Saxon/Simon Snorkel SS263	HP	1985	C07 Greenock
C316HGB	Bedford TK	Fulton & Wylie Fire Warrior	WrL	1985	E17 Biggar
C317HGB	Bedford TK	Fulton & Wylie Fire Warrior	WrL	1985	D28 Lamlash
C318HGB	Bedford TK	Fulton & Wylie Fire Warrior	WrL	1985	D14 Dalmellington
C812JGB	Scania 92	Angloco/Metz DL30	TL	1986	D02 Kilmarnock
C430KDS	Dodge G13T	Fulton & Wylie	WrL	1986	F25 Cove
C431KDS	Dodge G13T	Fulton & Wylie	WrL	1986	F06 Helensburgh
C432KDS	Dodge G13T	Fulton & Wylie	WrL	1986	D04 Ardrossaen
C433KDS	Dodge G16	F&W/Simon Snorkel SS263	HP	1986	D01 Ayr
C434KDS	Dodge G13T	Fulton & Wylie	WrL	1986	F14 Arrochar
D329LCS	Ford Transit	Fulton & Wylie	VSU	1986	North Reserve
D330LCS	Ford Transit	Fulton & Wylie	VSU	1986	North Reserve
D486PGD	Scania 82	Fulton & Wylie	WrL	1986	F16 Tobermory
D192PGD	Scania 82	Fulton & Wylie	WrL	1986	C11 Gourock
D193PGD	Scania 82	Fulton & Wylie	WrL	1986	D16 Cumnock
D194PGD	Scania 82	Fulton & Wylie	WrL	1986	D16 Cumnock
D195PGD	Scania 82	Fulton & Wylie	WrL	1986	E13 Strathaven
D196PGD	Scania 82	Fulton & Wylie	WrL	1986	F23 Rothesay
D377RGG	Ford Transit	Strathclyde FB	RRU	1987	C05 Renfrew
D549RGG	Scania 92	Angloco/Metz DL30	TL	1987	C04 Paisley
D999SSU	Scania 92	Fulton & Wylie	WrL	1987	F11 Stepps
E818VYS	Renault-Dodge G08	Scott	CU	1987	C01 Johnstone
E704WGB	Scania 92	Angloco/Metz DL30	TL	1987	F01 Clydebank
E991WNS	Scania 92	Angloco/Metz DL30	TL	1987	B05 Polmadie
E140XDS	Scania 82	Fulton & Wylie	WrL	1987	D27 Brodick
E141XDS	Scania 82	Fulton & Wylie	WrL	1987	E16 Abingdon
E142XDS	Scania 82	Fulton & Wylie	WrL	1987	D12 Girvan
E143XDS	Scania 82	Fulton & Wylie	WrL	1987	D20 Stewarton
E144XDS	Scania 82	Fulton & Wylie	WrL	1987	C06 Port Glasgow
E145XDS	Scania 82	Fulton & Wylie	WrL	1987	D22 Beith
E146XDS	Scania 82	Fulton & Wylie	WrL	1987	E15 Douglas
E147XDS	Scania 82	Fulton & Wylie	WrL	1987	D18 Mauchline
E148XDS	Scania 82	Fulton & Wylie	WrL	1987	D23 Kilbirnie
E460SSD	Ford Transit	Strathclyde FB	RRU	1988	E02 Motherwell
E461SSD	Ford Transit	Strathclyde FB	RRU	1988	F07 Dumbarton
E385XGE	Ford Transit	Strathclyde FB	VSU	1988	F65 Port Charlotte V, Isley
E695YNS	Scania 92	Saxon/Simon Snorkel SS263	HP	1988	E04 Coatbridge
F181FHS	Scania 82M	Angloco	WrL	1988	F07 Dumbarton
F182FHS	Scania 82M	Angloco	WrL	1988	C07 Greenock
F183FHS	Scania 82M	Angloco	WrL	1988	E06 Lanark
F184FHS	Scania 82M	Angloco	WrL	1988	C01 Johnstone
F185FHS	Scania 82M	Angloco	WrL	1988	F22 Inveraray
F186FHS	Scania 82M	Angloco	WrL	1988	E04 Coatbridge
F187FHS	Scania 82M	Angloco	WrL	1988	F12 Kilsyth
F188FHS	Scania 82M	Angloco	WrL	1988	D01 Ayr
F189FHS	Scania 82M	Angloco	WrL	1988	E06 Lanark
F190FHS	Scania 82M	Angloco	WrL	1988	C11 Gourock
E204FHS	Scania P93M-250	Angloco/Metz DL30	TL	1988	B07 Parkhead, Glasgow
F998HDS	Scania 93	Fulton & Wylie	WrC	1989	Training Centre
F998HDS	Scania 93	Fulton & Wylie	WrC	1989	Training Centre
F118HHS	Land Rover 110	Strathclyde FB	TV	1989	E15 Douglas
F83HNS	Renault-Dodge G08	Fulton & Wylie	ST	1989	A01 Cowcaddens, Glasgow
F898KYS	Mercedes-Benz 814D	Scott	TSU	1989	B02 Govan, Glasgow
F902JSU	Volvo FL6.17	Fulton & Wylie	WrL	1989	F15 Oban
F903JSU	Volvo FL6.17	Fulton & Wylie	WrL	1989	F14 Arrachar
F125LGG	Volvo FL6.17	Fulton & Wylie	WrL	1989	F24 Dunoon
F126LGG	Volvo FL6.17	Fulton & Wylie	WrL	1989	E11 Shotts
F268WCS	Ford Transit	Strathclyde FB	RRU	1989	E14 Lesmahagow
G531PGE	Scania 93-210	Fulton & Wylie	WrL	1989	A04 North West Glasgow
G532PGE	Scania 93-210	Fulton & Wylie	WrL	1989	A04 North West Glasgow
G533PGE	Scania 93-210	Fulton & Wylie	WrL	1989	A07 West Glasgow
G534PGE	Scania 93-210	Fulton & Wylie	WrL	1989	A07 West Glasgow
G535PGE	Scania 93-210	Fulton & Wylie	WrL	1989	F01 Clydebank

Based at Ayr is the hydraulic platform C433KDS of Strathclyde Fire Brigade. The vehicle is based on a Dodge chassis with cab and bodywork by Fulton & Wylie. *Keith Richardson*

Pictured when new, Strathclyde M58FYS, a Scania P113 with Bronto Skylift 32 metre booms fitted with Angloco bodywork. *Angloco Limited*

G536PGE	Scania 93-210	Fulton & Wylie	WrL	1989	F01 Clydebank
G537PGE	Scania 93-210	Fulton & Wylie	WrL	1989	F06 Helensburgh
G538PGE	Scania 93-210	Fulton & Wylie	WrL	1989	D24 Largs
G539PGE	Scania 93-210	Fulton & Wylie	WrL	1989	D26 Millport
G540PGE	Scania 93-210	Fulton & Wylie	WrL	1989	E04 Coatbridge
G839DCS	Ford Transit	Strathclyde FB	RRU	1990	F22 Inveraray
G495SYS	Scania 93M-210	Strathclyde FB	HRV	1990	A02 Easterhouse, Glasgow
H438WGG	Scania 93M-210		PM	1990	Training Centre
H439WGG	Scania 93M-210		PM	1990	Training Centre
H92YUS	Scania 93M-210	Fulton & Wylie	WrL	1990	F05 Milngavie
H93YUS	Scania 93M-210	Fulton & Wylie	WrL	1990	D04 Ardrossen
H94YUS	Scania 93M-210	Fulton & Wylie	WrL	1990	Training Center
H95YUS	Scania 93M-210	Fulton & Wylie	WrL	1990	A01 Cowcaddens, Glasgow
H96YUS	Scania 93M-210	Fulton & Wylie	WrL	1990	A01 Cowcaddens, Glasgow
H97YUS	Scania 93M-210	Fulton & Wylie	WrL	1990	B08 Calton, Glasgow
H98YUS	Scania 93M-210	Fulton & Wylie	WrL	1990	B08 Calton, Glasgow
H515CGD	Mercedes-Benz 814D	Fulton & Wylie	FoT/ST	1991	F01 Clydebank
H515CGD	Mercedes-Benz 814D	Fulton & Wylie	FoT/ST	1991	D05 Dreghorn
H515CGD	Mercedes-Benz 814D	Fulton & Wylie	FoT/ST	1991	B05 Polmadie
H101YUS	Scania 93M-210	Emergency One	WrL	1991	C05 Refrew
H102YUS	Scania 93M-210	Emergency One	WrL	1991	A05 Knightswood, Glasgow
H103YUS	Scania 93M-210	Emergency One	WrL	1991	B02 Govan
H104YUS	Scania 93M-210	Emergency One	WrL	1991	B02 Govan
H105YUS	Scania 93M-210	Emergency One	WrL	1991	B03 Pollock, Glasgow
H106YUS	Scania 93M-210	Emergency One	WrL	1991	B03 Pollock, Glasgow
H107YUS	Scania 93M-210	Emergency One	WrL	1991	C01 Johnstone
H108YUS	Scania 93M-210	Emergency One	WrL	1991	C02 Barrhead
J162GUS	Scania 93M-210	Emergency One	WrL	1991	C03 Clarkston
J163GUS	Scania 93M-210	Emergency One	WrL	1991	D01 Ayr
J164GUS	Scania 93M-210	Emergency One	WrL	1991	C06 Port Glasgow
J165GUS	Scania 93M-210	Emergency One	WrL	1991	E01 Hamilton
J166GUS	Scania 93M-210	Emergency One	WrL	1991	E01 Hamilton
J167GUS	Scania 93M-210	Emergency One	WrL	1992	Training Centre
J168GUS	Scania 93M-210	Emergency One	WrL	1991	A03 Springburn, Glasgow
J169GUS	Scania 93M-210	Emergency One	WrL	1992	A03 Springburn, Glasgow
J170GUS	Scania 93M-210	Emergency One	WrL	1991	F02 Bishopbriggs
J171GUS	Scania 93M-210	Emergency One	WrL	1991	E05 East Kilbride
J172GUS	Scania 93M-210	Emergency One	WrL	1991	E05 East Kilbride
J173GUS	Scania 93M-210	Emergency One	WrL	1991	F04 Cumberland
J174GUS	Scania 93M-210	Emergency One	WrL	1991	F04 Cumberland
J175GUS	Scania 93M-210	Emergency One	WrL	1992	B05 Polmadie
J176GUS	Scania 93M-210	Emergency One	WrL	1991	B05 Polmadie
J603HGB	Mercedes-Benz 410D	Emergency One	RRU	1991	A05 Knightswood, Glasgow
K371MYS	Scania G93M-210	Emergency One	WrL	1992	C07 Greenock
K372MYS	Scania G93M-210	Emergency One	WrL	1992	D01 Ayr
K373MYS	Scania G93M-210	Emergency One	WrL	1992	D03 Irvine
K374MYS	Scania G93M-210	Emergency One	WrL	1992	C04 Paisley
K375MYS	Scania G93M-210	Emergency One	WrL	1992	D05 Dreghorn
K376MYS	Scania G93M-210	Emergency One	WrL	1992	C04 Paisley

Strathclyde Fire Brigade now have over 120 Scania appliances with water tender ladders the most common type. Dreghorn have one appliance from the 1992 delivery and it is pictured while attending an incident.
Keith Richardson

K377MYS	Scania P113M-310	Angloco/Bronto Skylift F30HDT	ALP	1992	A01 Cowcaddens, Glasgow
K661OUS	Volvo FL6.18	Emergency One	WrL	1993	D29 Troon
K662OUS	Volvo FL6.18	Emergency One	WrL	1993	D21 Dalry
K663OUS	Volvo FL6.18	Emergency One	WrL	1993	F17 Bowmore
K664OUS	Volvo FL6.18	Emergency One	WrL	1993	F26 Garelochhead
K665OUS	Volvo FL6.18	Emergency One	WrL	1993	F13 Balloch
K667OUS	Volvo FL6.18	Emergency One	WrL	1993	F20 Campeltown
K668OUS	Volvo FL6.18	Emergency One	WrL	1993	E18 Carluke
K669OUS	Volvo FL6.18	Emergency One	WrL	1993	E12 Larkhall
K670OUS	Volvo FL6.18	Emergency One	WrL	1993	D15 New Cumnock
K529RDS	Steyr Pinzgauer 718K	Emergency One	L6P	1993	F16 Tobermory
L712UGA	Scania P93ML	Emergency One	WrL	1993	F03 Kirkintilloch
L713UGA	Scania P93ML	Emergency One	WrL	1993	B06 Cambuslang, Glasgow
L714UGA	Scania P93ML	Emergency One	WrL	1993	B06 Cambuslang, Glasgow
L715UGA	Scania P93ML	Emergency One	WrL	1993	E02 Motherwell
L716UGA	Scania P93ML	Emergency One	WrL	1993	E02 Motherwell
L717UGA	Scania P93ML	Emergency One	WrL	1993	A02 Easterhouse, Glasgow
L718UGA	Scania P93ML	Emergency One	WrL	1993	A02 Easterhouse, Glasgow
L719UGA	Scania P93ML	Emergency One	WrL	1993	D02 Kilmarnock
L720UGA	Scania P93ML	Emergency One	WrL	1993	D02 Kilmarnock
L721UGA	Scania P93ML	Emergency One	WrL	1993	B04 Castlemilk, Glasgow
L722UGA	Scania P93ML	Emergency One	WrL	1993	E03 Bellshill
L723UGA	Scania P93ML	Emergency One	WrL	1993	B04 Castlemilk, Glasgow
L724UGA	Scania P93ML	Emergency One	WrL	1993	F07 Dumbarton
L725UGA	Scania P93ML	Emergency One	WrL	1993	C07 Greenock
L726UGA	Scania N113DRB	Leicester Carriage	CU	1993	
L727UGA	Scania P93ML	Emergency One	WrL	1993	B07 Parkhead, Glasgow
L984VHF	Volkswagen LT50	Dependable	VSU	1993	D53 Corriecravie V
L806USU	Mercedes-Benz 410D	Emergency One	RRU	1993	D01 Ayr
L155XGE	Mercedes-Benz 418D	Emergency One	RRU	1994	C05 Renfrew
L162XGE	Iveco Daily 49.10	Emergency One	VSU	1994	
L163XGE	Iveco Daily 49.10	Emergency One	VSU	1994	F65 Port Charlotte V, Islay
L164YCS	Iveco Daily 49.10	Emergency One	VSU	1994	Lochgoilhead V
L165XGE	Iveco Daily 49.10	Emergency One	VSU	1994	
M902DDS	Scania P93ML-220	Emergency One	WrL	1994	F01 Clydebank
M903DDS	Scania P93ML-220	Emergency One	WrL	1994	F01 Clydebank
M904DDS	Scania P93ML-220	Emergency One	WrL	1994	E04 Coatbridge
M905DDS	Scania P93ML-220	Emergency One	WrL	1994	E04 Coatbridge
M906DDS	Scania P93ML-220	Emergency One	WrL	1994	A07 Yorkhill
M907DDS	Scania P93ML-220	Emergency One	WrL	1994	A07 Yorkhill
M908DDS	Scania P93ML-220	Emergency One	WrL	1994	
M909DDS	Scania P93ML-220	Emergency One	WrL	1994	F06 Helensburgh
M910DDS	Scania P93ML-220	Emergency One	WrL	1994	Training School
M911DDS	Scania P93ML-220	Emergency One	WrL	1994	D01 Ayr
M912DDS	Scania P93ML-220	Emergency One	WrL	1994	D01 Ayr
M913DDS	Scania P93ML-220	Emergency One	WrL	1994	E06 Lanark
M355DSU	Volvo FL6.18	Emergency One	WrL	1995	E17 Biggar
M356DSU	Volvo FL6.18	Emergency One	WrL	1995	F14 Arrochar
M357DSU	Volvo FL6.18	Emergency One	WrL	1995	D14 Dalmellington
M58FYS	Scania P113H-320	Angloco/Bronto Skylift F32HDT	ALP	1995	A01 Cowcaddens, Glasgow
	Steyr Pinzgauer 718K	Emergency One	L6P	1993	F69 Craighouse V, Jura
	Glencoe Hydrostatic	Glencoe	ATV	1995	E15 Douglas
N822JSU	Scania P93ML-220	Emergency One	WrL	1995	
N823JSU	Scania P93ML-220	Emergency One	WrL	1995	
N824JSU	Scania P93ML-220	Emergency One	WrL	1995	
N825JSU	Scania P93ML-220	Emergency One	WrL	1995	
N826JSU	Scania P93ML-220	Emergency One	WrL	1995	
N827JSU	Scania P93ML-220	Emergency One	WrL	1995	
N828JSU	Scania P93ML-220	Emergency One	WrL	1995	
N829JSU	Scania P93ML-220	Emergency One	WrL	1995	
N830JSU	Scania P93ML-220	Emergency One	WrL	1995	
N831JSU	Scania P93ML-220	Emergency One	WrL	1995	
N832JSU	Scania P93ML-220	Emergency One	WrL	1995	
N833JSU	Scania P93ML-220	Emergency One	WrL	1995	
N834JSU	Scania P93ML-220	Emergency One	WrL	1995	
N835JSU	Scania P93ML-220	Emergency One	WrL	1995	
N836JSU	Scania P93ML-220	Emergency One	WrL	1995	
N298NGG	Mercedes-Benz Sprinter 412D	Emergency One	RRU	1996	F07 Dunbarton

SUFFOLK FIRE SERVICE

Suffolk Fire Service, Colchester Road, Ipswich IP4 4SS

A00018	CDX922Y	Bedford TK	HCB-Angus	WrL	1983	Workshops
A00019	CDX923Y	Bedford TK	HCB-Angus	WrL	1983	Training School
A00020	DDX295Y	Bedford TK	HCB-Angus	WrL	1983	Training School
A00021	DDX296Y	Bedford TK	HCB-Angus	WrL	1983	Workshops
A00022	DDX297Y	Bedford TK	HCB-Angus	WrL	1983	Workshops
A00024	DDX299Y	Bedford TK	HCB-Angus	WrL	1983	Workshops
A00025	DDX300Y	Bedford TK	HCB-Angus	WrL	1983	Research & Development
A00026	C875TRT	Bedford TK	HCB-Angus	WrL	1985	07 Orford
A00027	C876TRT	Bedford TK	HCB-Angus	WrL	1985	11 Wrentham
A00028	C882URT	Bedford TK	HCB-Angus	WrL	1986	32 Ixworth
A00029	C883URT	Bedford TK	HCB-Angus	PRL	1986	18 Stradbroke
A00030	C884URT	Bedford TK	HCB-Angus	WrL	1986	17 Clifton Rd, Ipswich
A00031	D831DPV	Bedford TK	HCB-Angus	WrT	1987	01 Colchester Road, Ipswich
A00032	D832DPV	Bedford TK	HCB-Angus	WrL	1987	29 Wickhambrook
A00033	F351RBJ	Volvo FL6.14	HCB-Angus	WrL	1989	15 Beccles
A00034	F352RBJ	Volvo FL6.14	HCB-Angus	WrL	1989	34 Mildenhall
A00035	F353RBJ	Volvo FL6.14	HCB-Angus	WrL	1989	22 Stowmarket
A00036	F354RBJ	Volvo FL6.14	HCB-Angus	WrL	1989	04 Holbrook
A00037	F355RBJ	Volvo FL6.14	HCB-Angus	WrL	1989	33 Brandon
A00038	F356RBJ	Volvo FL6.14	HCB-Angus	WrL	1989	12 Southwold
A00039	F357RBJ	Volvo FL6.14	HCB-Angus	WrL	1989	26 Long Melford
A00040	F358RBJ	Volvo FL6.14	HCB-Angus	WrL	1989	20 Debenham
A00041	F359RBJ	Volvo FL6.14	HCB-Angus	WrL	1989	09 Leiston
A00042	F360RBJ	Volvo FL6.14	HCB-Angus	WrL	1989	31 Elmswell

An interesting history is attached to Suffolks S00001(VDV143X). It was new in 1982 to Devon Fire Service fitted with a refurbished 1965 Merryweather Turntable Ladder. When retired by Devon, Suffolk acquired the vehicle and built this interesting bodywork for it in the brigade workshops for use as a Breathing Apparatus Maintenance Unit. *Edmund Gray*

A00043	G611BGV	Volvo FL6.14	HCB-Angus	WrL	1990	23 Hadleigh
A00044	G612BGV	Volvo FL6.14	HCB-Angus	WrL	1990	10 Saxmundham
A00045	G613BGV	Volvo FL6.14	HCB-Angus	WrL	1990	13 Halesworth
A00046	G614BGV	Volvo FL6.14	HCB-Angus	WrL	1990	14 Bungay
A00047	G615BGV	Volvo FL6.14	HCB-Angus	WrL	1990	06 Felixstowe
A00048	J156LDX	Volvo FL6.14	Fulton & Wylie	WrL	1991	28 Haverhill
A00049	J157LDX	Volvo FL6.14	Fulton & Wylie	WrL	1991	35 Newmarket
A00050	J158LDX	Volvo FL6.14	Fulton & Wylie	WrL	1991	28 Haverhill
A00051	J159LDX	Volvo FL6.14	Fulton & Wylie	WrL	1991	16 Lowestoft
A00052	J160LDX	Volvo FL6.14	Fulton & Wylie	WrL	1991	06 Felixstowe
A00053	J161LDX	Volvo FL6.14	Fulton & Wylie	WrL	1991	25 Sudbury
A00054	J162LDX	Volvo FL6.14	Fulton & Wylie	WrL	1991	25 Sudbury
A00055	J163LDX	Volvo FL6.14	Fulton & Wylie	WrL	1991	35 Newmarket
A00056	J164LDX	Volvo FL6.14	Fulton & Wylie	WrL	1991	19 Eye
A00057	J165LDX	Volvo FL6.14	Fulton & Wylie	WrL	1991	24 Woodbridge
A00058	L691WRT	Volvo FL6.14	Excalibur CBK	WrL	1993	01 Colchester Road, Ipswich
A00059	L692WRT	Volvo FL6.14	Excalibur CBK	WrL	1993	03 Princes Street, Ipswich
A00060	L693WRT	Volvo FL6.14	Excalibur CBK	WrL	1993	03 Princes Street, Ipswich
A00061	L694WRT	Volvo FL6.14	Excalibur CBK	WrL	1993	30 Bury St Edmunds
A00062	L695WRT	Volvo FL6.14	Excalibur CBK	WrL	1993	30 Bury St Edmunds
A00063	M901DGV	Volvo FL6.14	Excalibur CBK	WrL	1994	01 Colchester Road, Ipswich
A00064	M902DGV	Volvo FL6.14	Excalibur CBK	WrL	1994	16 Lowestoft
A00065	M903DGV	Volvo FL6.14	Excalibur CBK	WrL	1994	21 Needham Market
A00066	M904DGV	Volvo FL6.14	Excalibur CBK	WrL	1994	24 Nayland
A00067	M921DGV	Volvo FL6.14	Excalibur CBK	PRL	1994	03 Princes Street, Ipswich
A00068	M922DGV	Volvo FL6.14	Excalibur CBK	PRL	1994	30 Bury St Edmunds
A00069	M923DGV	Volvo FL6.14	Excalibur CBK	PRL	1994	16 Lowestoft
A00070	P953NPV	Volvo FL6.14	Excalibur CBK	PRL	1994	35 Newmarket
A00071	P954NPV	Volvo FL6.14	Excalibur CBK	PRL	1994	28 Haverhill
A00072	P955NPV	Volvo FL6.14	Excalibur CBK	PRL	1994	06 Felixstowe
D00001	Pod	Suffolk FS	Water Carrier	WrC	1994	16 Lowestoft
D00002	Pod	Suffolk FS	Chemical Incident Unit	CIU	1994	01 Colchester Road, Ipswich
D00003	Pod	Suffolk FS	Meals Unit	CaV	1994	30 Bury St Edmunds
D00004	Pod	Suffolk FS/Welford	General Purpose	GP	1994	01 Colchester Road, Ipswich
D00005	Pod	Suffolk FS/Welford	General Purpose	GP	1994	16 Lowestoft
D00006	Pod	Suffolk FS/Welford	General Purpose	GP	1994	30 Bury St Edmunds
D00007	Pod	Suffolk FS	Hose Layer	HL	1994	30 Bury St Edmunds
D00008	Pod	Suffolk FS	Foam Tender	FoT	19	01 Colchester Road, Ipswich
D00009	Pod	Suffolk FS	Foam Tender	FoT	19	30 Bury St Edmunds
D00010	Pod	Suffolk FS	Foam Tender	FoT	19	16 Lowestoft
D00011	Pod	Suffolk FS	Foam Tender	FoT	19	06 Felixstowe
D00012	Pod	Suffolk FS	BA Training	BA(T)	19	01 Colchester Road, Ipswich
D00013	Pod	Suffolk FS	BA Training	BA(T)	19	30 Bury St Edmunds
D00014	Pod	Suffolk FS	BA Training	BA(T)	19	16 Lowestoft
D00015	Pod	Suffolk FS	Foam Tender	FoT	19	28 Haverhill
D00016	Pod	Suffolk FS	Communcations Unit	CU	1993	01 Colchester Road, Ipswich
R00001	YPV380S	Bedford TK	Powell Duffryn Rolonof	PM	1978	16 Lowestoft
R00002	HGV218V	Bedford TK1260	Powell Duffryn Rolonof	PM	1979	30 Bury St Edmunds
R00003	JBJ60V	Bedford TK	Powell Duffryn Rolonof	PM	1980	06 Felixstowe
R00004	C273TPV	Bedford TK	Powell Duffryn Rolonof	PM	1986	01 Colchester Road, Ipswich
R00005	L627YDX	Volvo FL6.14	Powell Duffryn Rolonof	PM	1994	01 Colchester Road, Ipswich
R00006	L628YDX	Volvo FL6.14	Powell Duffryn Rolonof	PM	1994	01 Colchester Road, Ipswich
S00002	VDV143X	Shelvoke & Drewry WX	G&T/Suffolk FS(1994)	BAC	1994	01 Colchester Road, Ipswich
S00003	J990SOL	Mercedes-Benz 1726	Carmichael/Magirus DL30	TL	1991	03 Princes Street, Ipswich
S00004	Boat	Dell Quay Inflatable	SuFire 1	FBt	19	Headquarters
S00005	Boat	Dell Quay Inflatable	SuFire 2	FBt	19	16 Lowestoft
S00006	B243NPF	Leyland Freighter 1716	Dairy Crest	WrC	1985	01 Colchester Road, Ipswich
S00007	D655HHV	Mercedes-Benz 1726	Dairy Crest	WrC	19	05 Framlingham
S00008	F518HRC	Mercedes-Benz 1726	Dairy Crest	WrC	19	25 Sudbury
S00009	K639SRT	Mercedes-Benz 1726	Angloco/Metz DLK30PLC	TL	1993	03 Prince Road, Ipswich
S00010	C155RRT	Dodge S56	Fire safety Display Unit	EXU	1985	Fire Safety
S00011	M194CRT	Mercedes-Benz Unimog U140L	Pesci/Suffolk FS	L4PM	1994	01 Colchester Road, Ipswich
S00012	M195CRT	Mercedes-Benz Unimog U140L	Pesci/Suffolk FS	L4PM	1994	30 Bury St Edmunds
S00013	M196CRT	Mercedes-Benz Unimog U140L	Pesci/Suffolk FS	L4PM	1994	16 Lowestoft

Note: S00001 was aquired from Devon FS in 1993, and rebuilt from a TL.
S00006-S00008 were orginally Milk Tankers.

SURREY FIRE & RESCUE SERVICE

Surrey Fire & Rescue Service, St David's, 70 Wray Park Road, Reigate RH2 0EJ

HPH946V	Ford A0609	Ford	CU	1980	35 Walton on Thames
YPA548Y	Dennis RS133	Dennis	WrL	1983	Training Centre
YPA549Y	Dennis RS133	Dennis	WrL	1983	Training Centre
B365JPM	Dennis RS133	Dennis	WrL	1984	Reserve
B366JPM	Dennis RS133	Dennis	WrL	1984	Training Centre
B367JPM	Dennis RS133	Dennis	WrL	1984	Driving School
B368JPM	Dennis RS133	Dennis	WrL	1984	24 Godalming
B369JPM	Dennis RS133	Dennis	WrL	1984	Training Centre
D421WPG	Mercedes-Benz 1222F	Polyma Fire Protection	WrL	1986	Reserve
D422WPG	Mercedes-Benz 1222F	Polyma Fire Protection	WrL	1987	16 Lingfield
D423WPG	Mercedes-Benz 1222F	Polyma Fire Protection	WrL	1987	Reserve
D424WPG	Mercedes-Benz 1222F	Polyma Fire Protection	WrL	1987	38 Chobham
D425WPG	Mercedes-Benz 1222F	Polyma Fire Protection	WrL	1987	23 Gomshall
D426WPG	Mercedes-Benz 1222F	Polyma Fire Protection	WrL	1987	27 Dunsfold
D635XPF	Mercedes-Benz 1222F	Polyma Fire Protection	WrL	1987	Reserve
D636XPF	Mercedes-Benz 1222F	Polyma Fire Protection	WrL	1987	Reserve
D637XPF	Mercedes-Benz 1222F	Polyma Fire Protection	WrL	1987	15 Oxted
D638XPF	Mercedes-Benz 1222F	Polyma Fire Protection	WrL	1987	Reserve
D639XPF	Mercedes-Benz 1222F	Polyma Fire Protection	WrL	1987	Reserve
D644XPF	Mercedes-Benz 1222F	Polyma Fire Protection	WrL	1987	Reserve
D645XPF	Mercedes-Benz 1222F	Polyma Fire Protection	WrL	1987	Reserve
D646XPF	Mercedes-Benz 1222F	Polyma Fire Protection	WrL	1987	32 Staines
D647XPF	Mercedes-Benz 1222F	Polyma Fire Protection	WrL	1987	Reserve
D648XPF	Mercedes-Benz 1222F	Polyma Fire Protection	WrL	1987	Reserve
E409DPJ	Mercedes-Benz 1222F	Polyma Fire Protection	WrL	1987	25 Haselmere

Surrey Fire and Rescue Service took delivery of a damage-control Unit in 1995. The Dennis SS and RS ranges are now out of production, therefore, M31KPA will be the last of this type for Surrey. Painshill fire station, near Cobham, is base to this John Dennis built appliance. *Edmund Gray*

The Surrey Fire and Rescue Service purchased five Mercedes-Benz 1726 water carriers in 1993, all with bodywork by Maidment Tankers finished by the brigades own workshops. K313OCR is stationed at Leatherhead. *Edmund Gray*

In late 1995 Surrey Fire And Rescue Service renewed all their aerial appliances when three Volvo FL10s with Bronto Skylift F32HDT booms were delivered from Angloco. The new aerial appliances have replaced three Ford chassied Hydraulic Platforms at Chertsey, Guildford and Leatherhead. *Keith Richardson*

E410DPJ	Mercedes-Benz 1222F	Polyma Fire Protection	WrL	1987	37 Camberley
E411DPJ	Mercedes-Benz 1222F	Polyma Fire Protection	WrL	1987	15 Oxted
E412DPJ	Mercedes-Benz 1222F	Polyma Fire Protection	WrL	1987	37 Camberley
F401MPE	Land Rover 110 Tdi	Land Rover HCPU/Surrey F&RS	L4T	1988	37 Camberley
F402MPE	Land Rover 110 Tdi	Land Rover HCPU/Surrey F&RS	L4T	1988	35 Walton on Thames
F403MPE	Land Rover 110 Tdi	Land Rover HCPU/Surrey F&RS	L4T	1988	25 Haselmere
F404MPE	Land Rover 110 Tdi	Land Rover HCPU/Surrey F&RS	L4T	1988	12 Dorking
F405MPE	Land Rover 110 Tdi	Land Rover HCPU/Surrey F&RS	L4T	1988	26 Farnham
G261APF	Dennis SS233	John Dennis	WrL	1989	22 Guildford
G262APF	Dennis SS233	John Dennis	WrL	1989	17 Epsom
G263APF	Dennis SS233	John Dennis	WrL	1989	28 Cranleigh
G264APF	Dennis SS233	John Dennis	WrL	1989	35 Walton on Thames
G265APF	Dennis SS233	John Dennis	WrL	1989	28 Cranleigh
G282APF	Ford Transit	Surrey F&RS	SRU	1989	11 Reigate
H962GPA	GMC Chevrolet 3500 4x4	Reynolds Boughton	RT	1990	14 Godstone
H963GPA	GMC Chevrolet 3500 4x4	Reynolds Boughton	RT	1990	33 Chertsey
H710JPL	Dennis SS233	John Dennis	WrL	1990	17 Epsom
H711JPL	Dennis SS233	John Dennis	WrL	1990	22 Guildford
H712JPL	Dennis SS233	John Dennis	WrL	1990	22 Guildford
H713JPL	Dennis SS233	John Dennis	WrL	1990	24 Godalming
H714JPL	Dennis SS233	John Dennis	WrL	1990	31 Egham
H715JPL	Dennis SS233	John Dennis	WrL	1990	35 Walton on Thames
J463MKB	Iveco-Ford Cargo	Iveco-Ford	DTV	1992	Driving School
K291XPF	GMC Chevrolet 3500 4x4	Reynolds Boughton	RT	1992	21 Painshill, Cobham
K652WPE	Range Rover 3.5V8	Surrey F&RS	CU	1992	Headquarters
K311OCR	Mercedes-Benz 1726	Maidment Tankers/Surrey FB	WrC	1993	14 Godstone
K312OCR	Mercedes-Benz 1726	Maidment Tankers/Surrey FB	WrC	1993	12 Dorking
K313OCR	Mercedes-Benz 1726	Maidment Tankers/Surrey FB	WrC	1993	13 Leatherhead
K314OCR	Mercedes-Benz 1726	Maidment Tankers/Surrey FB	WrC	1993	26 Farnham
K315OCR	Mercedes-Benz 1726	Maidment Tankers/Surrey FB	WrC	1993	37 Camberley
L873CPC	Dennis Rapier TF202	John Dennis	WrL	1993	11 Reigate
L874CPC	Dennis Rapier TF202	John Dennis	WrL	1993	11 Reigate
L875CPC	Dennis Rapier TF202	John Dennis	WrL	1993	36 Esher
L876CPC	Dennis Rapier TF202	John Dennis	WrL	1993	14 Godstone
L877CPC	Dennis Rapier TF202	John Dennis	WrL	1993	13 Leatherhead
L878CPC	Dennis Rapier TF202	John Dennis	WrL	1993	21 Painshill, Cobham
L879CPC	Dennis Rapier TF202	John Dennis	WrL	1993	12 Dorking
L880CPC	Dennis Rapier TF202	John Dennis	WrL	1993	34 Sunbury
L881CPC	Dennis Rapier TF202	John Dennis	WrL	1993	29 Woking
L882CPC	Dennis Rapier TF202	John Dennis	WrL	1993	29 Woking
L883CPC	Dennis Rapier TF202	John Dennis	WrL	1993	33 Chertsey
L884CPC	Dennis Rapier TF202	John Dennis	WrL	1993	33 Chertsey
L885CPC	Dennis Rapier TF202	John Dennis	L4T	1993	26 Farnham
M31KPA	Dennis SS241	John Dennis	ST	1995	21 Painshill, Cobham
N597GLF	Volvo FL10	Angloco/Bronto Skylift F32HDT	ALP	1995	33 Chertsey
N598GLF	Volvo FL10	Angloco/Bronto Skylift F32HDT	ALP	1995	22 Guildford
N599GLF	Volvo FL10	Angloco/Bronto Skylift F32HDT	ALP	1995	13 Leatherhead
Boat	Dateline Diver 5.5m	Inflatable Boat & Trailer	FBt	1993	35 Walton on Thames

Five new Land Rover 110s with Land Rovers own High Capacity Pick Up bodies were modified in the Surrey workshops in 1988. The modifications included adding a water tank and hose reels and the resulting appliances received the L4T designation. Some of the modifications are partially visible in this view of F404MPE.
Edmund Gray

TAYSIDE FIRE BRIGADE

Tayside Region Fire Brigade, Blackness Road, Dundee DD1 5PA

	MSR516P	Dodge K1113	HCB-Angus/Tayside FB	FoT/ST	1976	A4 Arbroath
	YTS208T	Dodge	Dodge	RT	1978	A Reserve
	FSP923W	Dodge G1313	HCB-Angus/Tayside FB	ET	1981	B Reserve
	FSP924W	Dodge G1313	HCB-Angus/Tayside FB	CU	1981	A1 Blackness Rd, Dundee
145	HSP130W	Dodge G1613	HCB-Angus/Simon Snorkel SS263	HP	1981	A2 MacAlpine Rd, Dundee
FT161	HSP131W	Dodge G1313	HCB-Angus	WrL	1981	Workshops
	HSP134W	Dodge G1313	HCB-Angus/Tayside FB	FoT/ST	1981	A11 Kingsway East, Dundee
FT163	OSR256Y	Dodge G13C	HCB-Angus	WrL	1983	Workshops
FT165	B804LSM	Dodge G13C	Fulton & Wylie	WrL	1984	B13 Kirkmichael Volunteers
FT164	B805LSM	Dodge G13C	Fulton & Wylie	WrL	1984	BReserve
FT169	C889YTS	Dodge G13C	Fulton & Wylie	WrL	1986	B14 Glenshee Volunteers
FT168	C890YTS	Dodge G13C	Fulton & Wylie	WrL	1986	Training Centre/Reserve
FT167	C891YTS	Dodge G13C	Fulton & Wylie	WrL	1986	A Reserve
FT166	C892YTS	Dodge G13C	Fulton & Wylie	WrL	1986	B Reserve
	D234ESR	Renault-Dodge G13C	Tayside FB	ET	1986	A2 MacAlpine Rd, Dundee
	D637ESL	Renault-Dodge G13C	Tayside FB	ET	1986	B1 Perth
FT173	D638ESL	Renault-Dodge G13C	Fulton & Wylie	WrL	1986	B3 Crieff
FT172	D639ESL	Renault-Dodge G13C	Fulton & Wylie	WrL	1986	A6 Brechin
FT171	D640ESL	Renault-Dodge G13C	Fulton & Wylie	WrL	1986	A12
FT170	D641ESL	Renault-Dodge G13C	Fulton & Wylie	WrL	1986	A Reserve
246	E689SEA	Scania P92M	Fulton & Wylie/Simon Snorkel SS263	HP	1987	B1 Perth
FT176	E356JSN	Renault-Dodge G13C	Mountain Range	WrL	1988	B12 Kinloch Rannoch Volunteers
FT175	E357JSN	Renault-Dodge G13C	Mountain Range	WrL	1988	B2 Auchterarder
FT174	E358JSN	Renault-Dodge G13C	Mountain Range	WrL	1988	A6 Brechin
	F680MTS	Mercedes-Benz 609D	Tayside FB	RT	1989	B7 Pitlochry
	F681MTS	Mercedes-Benz 609D	Tayside FB	RT	1989	B8 Kinross

Before the Dodge G series, the K series was a popular choice for fire appliances though very few survive with UK brigades into 1996. One that does is MSR516P of the Tayside Fire Brigade. Originally a Water Tender Ladder appliance in 1976, it was rebuilt at workshops as a Foam Salvage Tender and is currently on the run at Arbroath. *Keith Grimes*

FT181	F528OES	Volvo FL6.14	Mountain Range	WrL	1989	A Reserve
FT180	F529OES	Volvo FL6.14	Mountain Range	WrL	1989	B9 Coupar Angus
FT179	F272OSP	Renault-Dodge G13C	Mountain Range	WrL	1989	B7 Pitlochry
FT178	F273OSP	Renault-Dodge G13C	Mountain Range	WrL	1989	B8 Kinross
FT177	F274OSP	Renault-Dodge G13C	Mountain Range	WrL	1989	B2 Auchterarder
	G431RTS	Mercedes-Benz 609D	Tayside FB	RT	1989	A5 Montrose
	G432RTS	Mercedes-Benz 609D	Tayside FB	RT	1989	A7 Forfar
	G538STS	Mercedes-Benz 1114	Powell Duffryn Rolonoff	PM	1990	A1 Blackness Rd, Dundee
	G539STS	Mercedes-Benz 1114	Powell Duffryn Rolonoff	PM	1990	A3 Monifieth
FT185	G821TSP	Volvo FL6.14	Excalibur CBK	WrL	1990	B10 Blairgowrie
FT184	G822TSP	Volvo FL6.14	Excalibur CBK	WrL	1990	B10 Blairgowrie
FT183	G823TSP	Volvo FL6.14	Excalibur CBK	WrL	1990	A9 Carnoustie
FT182	G824TSP	Volvo FL6.14	Excalibur CBK	WrL	1990	B6 Brechin
	Pod	Penman/Tayside FB	BA Support Unit	BASU	1990	A3 Monifieth
	Pod	Penman	B A Training Unit	BA(T)	1990	A1 Blackness Rd, Dundee
	Pod	Penman	Flat-bed lorry	FBL	1990	A1 Blackness Rd, Dundee
	Pod	Penman/Tayside FB	Fire Prevention Unit	FPU	1990	A1 Blackness Rd, Dundee
	Pod	Penman/Tayside FB	Operations Support Unit	OSU	1990	A3 Monifieth
FT199	H856YSR	Volvo FL6.14	Emergency One	WrL	1991	B3 Crieff
FT188	H857YSR	Volvo FL6.14	Emergency One	WrL	1991	B11 Alyth
FT187	H858YSR	Volvo FL6.14	Emergency One	WrL	1991	A3 Monifieth
FT186	H859YSR	Volvo FL6.14	Emergency One	WrL	1991	A4 Arbroath
FT193	J389ESN	Volvo FL6.14	Emergency One	WrL	1991	A7 Forfar
FT192	J390ESN	Volvo FL6.14	Emergency One	WrL	1991	A7 Forfar
FT191	J391ESN	Volvo FL6.14	Emergency One	WrL	1991	A5 Montrose
FT198	J392ESN	Volvo FL6.14	Emergency One	WrL	1991	A5 Montrose
FT197	K37LES	Volvo FL6.14	Emergency One	WrL	1993	B4 Comrie
FT196	K38LES	Volvo FL6.14	Emergency One	WrL	1993	B5 Dunkeld
FT195	K39LES	Volvo FL6.14	Emergency One	WrL	1993	A8 Kirriemuir
FT194	K41LES	Volvo FL6.14	Emergency One	WrL	1993	A8 Kirriemuir
1695	K497MSR	Volvo FL10	Bedwas/Simon Snorkel ST290S	ALP	1993	A1 Blackness Rd, Dundee
FT224	L511SSN	Volvo FL6.14	Emergency One	WrL	1993	B1 Perth
FT225	L512SSN	Volvo FL6.14	Emergency One	WrL	1993	B1 Perth
FT226	L513SSN	Volvo FL6.14	Emergency One	WrL	1993	A11 Kingsway East, Dundee
FT227	L514SSN	Volvo FL6.14	Emergency One	WrL	1993	A11 Kingsway East, Dundee
FT229		Crayford Argocat 8x8	Crayford	ATV	1994	B1 Perth
FT230	Trailer	Crayford Engineering	Trailer for Argocat	Tr	1994	B1 Perth
FT235	M480WTS	Volvo FL6.14	Emergency One	WrL	1995	A2 MacAlpine Rd, Dundee
FT236	M481WTS	Volvo FL6.14	Emergency One	WrL	1995	A2 MacAlpine Rd, Dundee
FT237	M482WTS	Volvo FL6.14	Emergency One	WrL	1995	A3 Moniteith
FT238	M483WTS	Volvo FL6.14	Emergency One	WrL	1995	A4 Arbroath
FT246	N454ESN	Scania G93M-230	Emergency One	WrL	1996	A1 Blackness Rd, Dundee
FT247	N455ESN	Scania G93M-230	Emergency One	WrL	1996	A1 Blackness Rd, Dundee

Note: MSR516P, FSP923W, FSP924W & HSP134W were all previously WrLs.

Previous Registrations:

J389ESN	J389JSN	J391ESN	J391JSN
J390ESN	J390JSN	J392ESN	J392JSN

Four Mercedes-Benz 609D chassis received bodywork built at the brigades own workshops in 1989. These appliances are Rescue Tenders based at Forfar, Kinross, Montrose and Pitlochry retained fire stations. G432RTS is the Forfar appliance and demonstrates the attractive bodywork on these appliances.
Keith Grimes

TYNE AND WEAR METROPOLITAN FIRE BRIGADE

Tyne and Wear County Fire Brigade, Pilgrim Street, Newcastle-upon-Tyne NE99 1HR

160	L172VTN	Volvo FL6.14	Excalibur CBK	RP	1993	N Sunderland Central
161	L173VTN	Volvo FL6.14	Excalibur CBK	PP	1993	A West Denton
163	L184VTN	Volvo FS7.18	John Dennis/Atlas AS130.1	ST	1993	T Hebburn
310	BTN47V	Dodge G7575	David Earl	PM	1980	Community Liaison Unit
319	EBB846W	Dennis Dominator SD130A	Angloco	CU	1980	A West Denton
345	A204YCU	Dennis DF133	Chubb	FoT	1983	V Gateshead
364	A726BTY	Dennis DF133	Carmichael/Magirus DL30	TL	1983	Reserve
371	B60FVK	Dennis DF133	Carmichael/Magirus DL30	TL	1984	M Fulwell, Sunderland North
373	B759GCN	Dennis SS133	Dennis/T&W MFB	BT	1984	F Walker, Newcastle
398	B767JTN	Dennis DF133	Carmichael/Magirus DL30	TL	1985	V Gateshead
408	C758OFT	Dennis SS133	Carmichael	P	1986	W Birtley
409	C759OFT	Dennis SS133	Carmichael	PL	1986	Training Centre
410	C760OFT	Dennis SS133	Carmichael	P	1986	Training Centre
411	C761OFT	Dennis SS133	Carmichael	P	1986	Training Centre
412	C762OFT	Dennis SS133	Carmichael	PL	1986	Training Centre
422	D783XRG	Dennis SS133	Fulton & Wylie	P	1987	Z Chopwell
423	D784XRG	Dennis SS133	Fulton & Wylie	PL	1987	Reserve
424	D785XRG	Dennis SS133	Fulton & Wylie	PL	1987	Reserve
425	D786XRG	Dennis SS133	Fulton & Wylie	P	1987	Reserve
426	D787XRG	Dennis SS133	Fulton & Wylie	PL	1987	Reserve
443	F843JBB	Volvo FL6.14	Fulton & Wylie	ET	1988	K South Shields
446	F844JBB	Volvo FL6.14	Fulton & Wylie	P	1988	Reserve
447	F845JBB	Volvo FL6.14	Fulton & Wylie	P	1988	O Grindon, Sunderland West
448	F846JBB	Volvo FL6.14	Fulton & Wylie	P	1988	S Washington
449	F847JBB	Volvo FL6.14	Fulton & Wylie	P	1988	Reserve
450	F848JBB	Volvo FL6.14	Fulton & Wylie	P	1988	R Tunstall, Sunderland South
469	G262RJR	Volvo FL6.14	Excalibur CBK	RP	1989	Reserve
470	G263RJR	Volvo FL6.14	Excalibur CBK	RP	1989	O Grindon, Sunderland West
471	G264RJR	Volvo FL6.14	Excalibur CBK	P	1989	N Sunderland Central
472	G265RJR	Volvo FL6.14	Excalibur CBK	P	1989	E Gosforth
473	G266RJR	Volvo FL6.14	Excalibur CBK	P	1989	Y Swalwell
483	G830PNX	GMC 3500 4x4	Reynolds Broughton	RT	1989	T Hebburn
499	Tinaea	Marshall Branson	Fire Boat	FBt	1990	F Walker, Newcastle
500	Vedra	Marshall Branson	Fire Boat	FBt	1990	N Sunderland Central
502	G935XFT	Volvo FL6.14	Excalibur CBK	P	1990	A West Denton
504	G932XFT	Volvo FL6.14	Excalibur CBK	P	1990	T Hebburn
506	G934XFT	Volvo FL6.14	Excalibur CBK	P	1990	M Fulwell, Sunderland North
508	G931XFT	Volvo FL6.14	Excalibur CBK	P	1990	G Wallsend
509	H588ACN	Volvo FL6.14	Excalibur CBK	P	1990	J North Shields, Tynemouth
510	H593ACN	Volvo FL6.14	Excalibur CBK	P	1990	B Benwell, Newcastle West
511	H589ACN	Volvo FL6.14	Excalibur CBK	RP	1990	B Benwell, Newcastle West
512	H590ACN	Volvo FL6.14	Excalibur CBK	P	1990	K South Shields
513	H591ACN	Volvo FL6.14	Excalibur CBK	P	1990	V Gateshead
514	H592ACN	Volvo FL6.14	Excalibur CBK	P	1990	D Newcastle Central
105	Trailer	Selby Coach Works	Fire safety Demonstration	ExU	1994	Community Liaison
128	D365YCP	Ford Transit	Ford Minibus	PCV	1986	City Challenge, Newcastle
129	L422YRG	Ford Transit	Ford Minibus	PCV	1993	T Hebburn
130	J376MBB	Ford Transit	Ford Minibus	PCV	1992	N Sunderland Central
131	H908YJR	Ford Transit	Ford Minibus	PCV	1991	City Challenge, Newcastle
142	FT7215	Dennis F12	Dennis	PE	1951	Museum, North Shields
143	EBB302	Leyland Cub	Leyland	PE	19	Museum, North Shields
164	L203ANL	Volvo FL6.14	Excalibur CBK	RP	1994	Y Swalwell
165	L205ANL	Volvo FL6.14	Excalibur CBK	RP	1994	E Gosforth
166	L206ANL	Volvo FL6.14	Excalibur CBK	RP	1994	T Hebburn
167	L202ANL	Volvo FL6.14	HCB-Angus	ET	1994	D Newcastle Central
168	L204ANL	Volvo FL6.14	Excalibur CBK	RP	1994	V Gateshead
206	M556KVK	Volvo FL6.14	Excalibur CBK	RP	1995	G Wallsend
229	N389RCN	Volvo FL6.14	Excalibur CBK	ET	1996	N Sunderland Central

The Fire Brigade Handbook

Although sharing the standard design of steel safety cab with the RS and SS ranges, the Dennis DF was a heavier duty chassis designed for specialist applications such as Tyne and Wear's 345 (A204YCU) at Gateshead. This Chubb-bodied machine is the brigades sole front line Foam Tender. The older Foam Tender and now reserve is EBB847W based on a Shelvoke & Drewery unit and is seen in Gateshead.
Karl Sillitoe/
Keith Grimes

Three GMC 3500 Rescue Tenders are based at Hebburn, Walker and Sunderland Central fire stations. When required, these appliances also tow trailers to incidents. Those based at Walker and Sunderland Central tow trailers with Marshall Branson fireboats. This view shows former Sunderland Central's 460 (F156SRE) recently sold. *Karl Sillitoe*

An unusual feature of Rescue Pumps specified by Tyne & Wear Metropolitan Fire Brigade, is a roof access ladder built into the appliance bodywork from the 1995 delivery. This is visible in this view of Tynemouth's new appliance 228 (N388RCN). In common with all other Volvo pumps delivered since 1989, it has Excalibur CBK bodywork. *Tony Wilson*

208	M558KVK	Volvo FL6.14	Excalibur CBK	RP	1995	F Walker, Newcastle
209	M559KVK	Volvo FL6.14	Excalibur CBK	RP	1995	S Washington
210	M561KVK	Volvo FL6.14	Excalibur CBK	RP	1995	W Birtley
224	N384RCN	Volvo FL6.14	Excalibur CBK	RP	1996	M Fulwell, Sunderland North
225	N385RCN	Volvo FL6.14	Excalibur CBK	RP	1996	K South Shields
226	N386RCN	Volvo FL6.14	Excalibur CBK	RP	1996	R Tunstall, Sunderland South
227	N387RCN	Volvo FL6.14	Excalibur CBK	RP	1996	D Newcastle Central
228	N388RCN	Volvo FL6.14	Excalibur CBK	RP	1996	J North Shields, Tynemouth
230	M304LBB	Iveco Ford Euro Fire 150E27	GB Fire/Magirus DLK18-12	TL	1995	F Walker, Newcastle
317	BTN48V	Dodge G7575	David Earl	PM	1980	Community Liaison
320	EBB847W	Shelvoke & Drewry WX	Chubb	FoT	1981	Reserve
339	PVK889Y	Dodge G08	David Earl	PM	1983	Community Liaison
362	A729BTY	Dennis SS133	Dennis	PL	1984	City Challenge, Newcastle
365	A645CJR	Ford Cargo	David Earl	PM	1984	Community Liaison
374	B434GRG	Dennis SS133	Dennis	PL	1985	City Challenge, Newcastle
403	Pod	David Earl	Community Liaison Unit	CLU	1980	Community Liaison
404	Pod	David Earl	Community Liaison Unit	CLU	1980	Community Liaison
405	Pod	David Earl	Community Liaison Unit	CLU	1980	Community Liaison
406	Pod	David Earl	Community Liaison Unit	CLU	1980	Community Liaison
431	E887ATY	Ford Transit	Ford Minibus	PCV	1987	Y Swalwell
492	Trailer	Indispension	Boat Transporter	BT	1990	N Sunderland Central
493	Trailer	Indispension	Boat Transporter	BT	1990	F Walker, Newcastle
522	Trailer	Tow-a-Van	Chemical Incident Unit	CIU	1989	T Hebburn
523	Trailer	Tow-a-Van	Building Collapse Unit	BCU	1989	T Hebburn
524	Trailer	Tow-a-Van	BA Support Unit	BASU	1989	Operations Department
541	Trailer	Tow-a-Van	Foam Equipment Unit	FEU	1991	T Hebburn

WARWICKSHIRE FIRE & RESCUE SERVICE

Warwickshire Fire & Rescue Service, Warwick Street, Leamington Spa CV32 5LH

BAT2	D576OOM	Freight Rover Sherpa 400	Warwickshire F&RS	BAT	1987	24 Atherstone
CU1	B388PWK	Bedford VAS5	Benson	CU	1984	29 Leamington Spa
GPT2	D39CWK	Leyland Roadrunner	Leyland	GPT	1987	Headquarters
GPT4	C843VDH	Bedford TL	Bedford	GPT	1985	20 Nuneaton
HP1	DVC274Y	Shelvoke & Drewry WY	HCB-Angus/Simon Snorkel SS263	HP	1984	29 Leamington Spa
HP4	D754EAC	Renault-Dodge G16	Saxon/Simon Snorkel SS263	HP	1986	20 Nuneaton
PERC2	M439ERW	Land Rover Defender 110 Tdi	Land Rover Station Wagon	PCV	1994	35 Stratford-upon-Avon
PERC5	M765KVC	Land Rover Defender 110 Tdi	Land Rover Station Wagon	PCV	1995	20 Nuneaton
PERC8	H765ERW	Leyland-DAF 400	Leyland-DAF	PCV	1991	Training School
PERC9	H766ERW	Leyland-DAF 400	Leyland-DAF	PCV	1991	29 Leamington Spa
PM1	G551XWK	Dennis DFS237	Multilift	PM	1989	22 Coleshill
PM2	G552XWK	Dennis DFS237	Multilift	PM	1989	20 Nuneaton
PM3	G553XWK	Dennis DFS237	Multilift	PM	1989	29 Leamington Spa
PM4	J904VRW	Dennis DFS237 6x2	Multilift	PM	1992	Driving Centre (29)
Pod1	Pod	Transliner	Emergency Tender	ET	1989	20 Nuneaton
Pod2	Pod	Transliner	Emergency Tender	ET	1989	29 Leamington Spa
Pod3	Pod	Transliner	Foam Tender	FoT	1991	22 Coleshill
Pod4	Pod	Whale Tankers	Water Carrier	WrC	1991	22 Coleshill
Pod5	Pod	Whale Tankers/WFRS	Water Carrier	WrC	1992	29 Leamington Spa
Pod6	Pod	Multilift	General Purpose Unit	GPU	1993	29 Leamington Spa
UV7	J397PHP	Land Rover Defender 110 Tdi	Land Rover	L4V	1991	26 Rugby
UV8	J398PHP	Land Rover Defender 110 Tdi	Land Rover	L4V	1991	29 Leamington Spa
UV9	A708JVC	Land Rover 110	Land Rover	L4V	1983	Headquarters
UV34	M618NWK	Land Rover Defender 110 Tdi	Land Rover	L4V	1995	22 Coleshill

Three new Dennis Sabre Pump Rescue Ladder appliances were delivered in 1995, part of the specification included Angus Sacol ladder gantries for the John Dennis bodywork. PRL68 (M68PWK) was photographed prior to going on the run at Atherstone. *Edmund Gray*

Prime Movers and Pods are used by Warwickshire for several specialist roles, all of the prime movers are Dennis DFS237s with Multilift pod handling equipment. The latest is of unconventional tri-axle layout with a crew cab and is used primarily for driver training. However a standard example (PM1, G551XWK) is illustrated here carrying Pod 1, built by Transliner. *Simon Pearce*

WRT7	C316WRW	Bedford TKG	HCB-Angus HSC	WrL	1985	Reserve
WRT15	C318WRW	Bedford TKG	HCB-Angus HSC	WrL	1985	24 Atherstone
WRT16	A853XOP	Bedford TKG	HCB-Angus CSV	WrT	1984	Reserve
WRT24	C320WRW	Bedford TKG	HCB-Angus HSC	WrL	1986	27 Kenilworth
WRT35	C322WRW	Bedford TKG	HCB-Angus HSC	WrL	1985	Fire Safety - Welephant
WRT37	A856XOP	Bedford TKG	HCB-Angus CSV	WrT	1984	Reserve
PRL46	D781GHP	Dennis SS137	Carmichael	WrT	1987	37 Alcester
PRL47	D782GHP	Dennis SS137	Carmichael	PRL	1987	25 Brinklow
PRL48	D783GHP	Dennis SS137	Carmichael	WrL	1987	33 Tysoe
PRL49	D784GHP	Dennis SS137	Carmichael	WrT	1987	36 Bidford-on-Avon
PRL50	F950PWK	Dennis SS237	Carmichael	WrL	1988	21 Bedworth
PRL51	F951PWK	Dennis SS135	Carmichael	WrT	1988	Training School
PRL52	F952PWK	Dennis SS237	Carmichael	PRL	1988	34 Shipston
PRL53	F953PWK	Dennis SS237	Carmichael	WrL	1988	32 Kineton
PRL54	G654XWK	Dennis SS237	Carmichael	WrL	1989	23 Polesworth
PRL55	G655XWK	Dennis SS237	Carmichael	WrL	1989	31 Fenney Compton
PRL56	G656XWK	Dennis SS237	Carmichael	WrL	1989	38 Studley
PRL57	G857DVC	Dennis SS237	Carmichael	WrL	1990	30 Southam
PRL59	G859DVC	Dennis SS237	Carmichael	WrL	1990	35 Stratford-upon-Avon
PRL60	J460SAC	Dennis SS237	Carmichael	PRL	1991	Reserve
PRL61	J461SAC	Dennis SS237	Carmichael	WrT	1991	Reserve
PRL62	K62DKV	Dennis SS237	John Dennis	WrL	1992	28 Warwick
PRL63	K63DKV	Dennis SS237	John Dennis	WrT	1992	Driving Centre 20
PRL64	K464KHP	Dennis Rapier TF202	John Dennis	PRL	1993	21 Bedworth
PRL65	L565RVC	Dennis SS237	John Dennis	PRL	1993	22 Coleshill
PRL66	L566RVC	Dennis SS237	John Dennis	PRL	1993	39 Henley-in-Arden
PRL67	M67PWK	Dennis Sabre TSD233	John Dennis	PRL	1995	20 Nuneaton
PRL68	M68PWK	Dennis Sabre TSD233	John Dennis	PRL	1995	24 Atherstone
PRL69	M69PWK	Dennis Sabre TSD233	John Dennis	PRL	1995	20 Nuneaton
PRL70	N170WRW	Dennis Sabre S411	John Dennis	PRL	1995	29 Leamington Spa
PRL71	N171WRW	Dennis Sabre S411	John Dennis	PRL	1995	26 Rugby
PRL72	N172WRW	Dennis Sabre S411	John Dennis	PRL	1995	29 Leamington Spa
PRL73	P73HHP	Dennis Sabre S411	John Dennis	PRL	1996	26 Rugby
PRL74	P74HHP	Dennis Sabre S411	John Dennis	PRL	1996	35 Stratford-upon-Avon

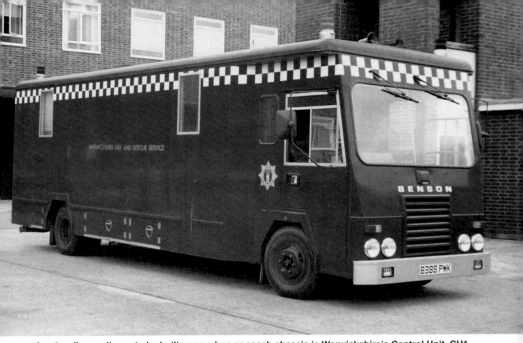

Another fire appliance to be built upon a bus or coach chassis is Warwickshire's Control Unit, CU1 (B388PWK). The Bedford VAS was primarily used as the basis for midicoaches. The specialist bodywork for this vehicle was by Benson. It is based at Headquarters, in Leamington Spa. *Simon Pearce*

Warwickshire has been applying high visibility markings onto its appliances recently, with various permutations of reflective yellow, red, white and silver stripes. Shown here is PRL62 (K62DKV), one of a pair of Dennis SS234s in the fleet. *Graham Hopwood*

WEST MIDLANDS FIRE SERVICE

West Midlands County Fire Brigade, Lancaster Circus, Queensway, Birmingham B4 7DE

		Shand Mason	Double Vertical Steam Pump	P	1901	Preserved at A3 Sutton Coldfield
234	YOX235K	Daimler Fleetline CRG6LX	MCW	ExU	1972	Community Relations
135	KVP178P	ERF 84PS	ERF/Simon Snorkel SS85	HP	1975	B3 Coventry (R)
086	Q68VOE	Shelvoke & Drewry WX	Carmichael (1987)	PRL	1977	Training Centre
085	Q69VOE	Shelvoke & Drewry WX	Carmichael (1987)	PRL	1977	Training Centre
089	Q70VOE	Shelvoke & Drewry WX	Carmichael (1987)	PRL	1977	Training Centre
300	POB982R	ERF 84CS	ERF/Simon Snorkel SS263	HP	1977	C7 Ladywood (R)
236	TVP849S	Leyland National 11351A/1R	Leyland National	ExU	1978	Community Relations
217	UVP98S	ERF 84PS	Angloco/Magirus DL30	TL	1978	D4 Halesowen (R)
114	DOJ113V	Dennis Dart SD504	Marshall Campagna	PCh	1980	Training Centre
130	DOJ115V	Dennis F131	Dennis	UV	1979	Uniform Stores
125	HOP539W	Shelvoke & Drewry WY	Angloco/Simon Snorkel SS263	HP	1980	B2 Sheldon
116	HOP548W	Dodge RG16		PCV	1981	A3 (Brigade Band)
009	LOX811X	Dennis RS133	Dennis	PRL	1981	B4 Canley (R)
018	LOX813X	Dennis RS133	Dennis	PRL	1981	Withdrawn
136	NOV870X	Shelvoke & Drewry WY	Angloco/Simon Snorkel SS263	HP	1982	B1 Solihull
007	A21WOB	Dennis RS133	Dennis	PRL	1983	C Training
015	A995WUK	Dennis RS133	Dennis	PRL	1983	A Reserve
030	A996WUK	Dennis RS133	Dennis	PRL	1983	Workshops
304	A954XOB	Dennis Delta DF1616 6x2	Multilift	PM	1984	C3 Smethwick
307	A957XOB	Dennis Delta DF1616	Multilift	PM	1984	E1 Walsall
004	B52BVP	Dennis RS133	Dennis	PRL	1984	Training Centre
005	B53BVP	Dennis RS133	Dennis	PRL	1984	Training Centre
017	B54BVP	Dennis RS133	Dennis	PRL	1984	Training Centre
022	B55BVP	Dennis RS133	Dennis	PRL	1984	Training Centre
033	B56BVP	Dennis RS133	Dennis	PRL	1984	C2 Harbourne (R)
045	B57BVP	Dennis RS133	Dennis	PRL	1984	Training Centre
054	B999BUK	Dennis RS133	Dennis	PRL	1984	C8 Billesley (R)
025	C126GOH	Dennis RS133	Dennis	PRL	1984	B6 Binley (R)
060	C127GOH	Dennis RS133	Dennis	PRL	1985	C5 Kings Norton (R)
062	C128GOH	Dennis RS133	Dennis	PRL	1985	B5 Foleshill (R)
066	C129GOH	Dennis RS133	Dennis	PRL	1985	B4 Canley (R)
069	C132GOH	Dennis RS133	Dennis	PRL	1985	D5 Stourbridge (R)
070	C133GOH	Dennis RS133	Dennis	PRL	1985	D5 Stourbridge (R)
067	C548HOE	Dennis RS133	Dennis	PRL	1985	E4 Aldridge (R)
068	C549HOE	Dennis RS133	Dennis	PRL	1985	E7 Bilston
115	C964JOK	Dennis DFD133	Carmichael/Magirus DL30	TL	1986	D2 Brierley Hill
003	D84NOX	Dennis RS133	Carmichael	PRL	1987	A3 Sutton Coldfield (R)
006	D85NOX	Dennis RS133	Carmichael	PRL	1987	A3 Sutton Coldfield (R)
049	D86NOX	Dennis RS133	Carmichael	PRL	1987	A2 Aston
061	D87NOX	Dennis RS133	Carmichael	PRL	1987	D6 Cradeley Heath
071	D88NOX	Dennis RS133	Carmichael	PRL	1987	B1 Solihull
079	D89NOX	Dennis RS133	Carmichael	PRL	1987	D7 Tipton
080	D90NOX	Dennis RS133	Carmichael	PRL	1987	E9 Wednesbury
082	D91NOX	Dennis RS133	Carmichael	PRL	1987	E6 Fallings Park
113	D584OOP	Dennis DFD133	Carmichael/Magirus DL30	TL	1987	A1 Central Birmingham
043	E769SOJ	Dennis RS133	Carmichael	PRL	1987	E4 Aldridge
031	E770SOJ	Dennis SS239	Carmichael	PRL	1987	D6 Cradeley Heath
050	E771SOJ	Dennis RS133	Carmichael	PRL	1987	D7 Tipton
157	E819SOJ	Freight Rover Sherpa	Freight Rover	PCV	1987	Training Centre
308	E823SOJ	Leyland-DAF Freighter 1617	Multilift	PM	1987	A5 Perry Bar
011	F505WVP	Dennis RS133	Mountain Range	PRL	1988	B4 Canley
012	F506WVP	Dennis RS133	Mountain Range	PRL	1988	Training Centre
016	F507WVP	Dennis RS133	Mountain Range	PRL	1988	E3 Willenhall
019	F508WVP	Dennis RS133	Mountain Range	PRL	1988	D2 Brierly Hill
028	F509WVP	Dennis RS133	Mountain Range	PRL	1988	C6 Northfield
034	F510WVP	Dennis RS133	Mountain Range	PRL	1988	A2 Aston
047	F511WVP	Dennis RS237	Mountain Range	PRL	1988	A7 Handsworth
074	F512WVP	Dennis RS237	Mountain Range	PRL	1988	D1 Oldbury
249	F514WVP	Freight Rover Sherpa	Freight Rover	PCV	1988	B3 Coventry
252	F515WVP	Freight Rover Sherpa	Freight Rover	PCV	1988	E5 Wolverhampton
117	F516WVP	Dodge G16	Gloster Saro/Simon Snorkel SS263	HP	1988	B3 Coventry
128	F854BOE	Land Rover 110	Land Rover	LR	1989	A3 Sutton Coldfield
129	F855BOE	Land Rover 110	Land Rover	LR	1989	D8 West Bromwich

With only a couple of exceptions, all pumping appliances with West Midlands Fire service are Dennis RS and SS models. Carmichael-bodied 103, H188MOK, is a SS237 example and was seen at a public muster day representing Walsall station. *Bill Murray*

024	G662FJW	Dennis RS133	Mountain Range	PRL	1988	A4 Erdington
032	G663FJW	Dennis RS133	Mountain Range	PRL	1988	B5 Foleshill
058	G664FJW	Dennis RS133	Mountain Range	PRL	1988	C2 Harbourne
064	G665FJW	Dennis RS133	Mountain Range	PRL	1988	Training Centre
065	G666FJW	Dennis RS133	Mountain Range	PRL	1988	A5 Perry Bar
075	G667FJW	Dennis RS133	Mountain Range	PRL	1988	C3 Smethwick
083	G668FJW	Dennis RS133	Mountain Range	PRL	1989	C8 Billesley
091	G669FJW	Dennis RS133	Mountain Range	PRL	1988	E6 Fallings Park
092	G670FJW	Dennis RS237	Saxon Sanbec	PRL	1990	A1 Central Birmingham
093	G671FJW	Dennis RS237	Saxon Sanbec	PRL	1990	A3 Sutton Coldfield
094	G672FJW	Dennis RS237	Saxon Sanbec	PRL	1990	B2 Sheldon
095	G673FJW	Dennis RS237	Saxon Sanbec	PRL	1990	C1 Highgate
096	G674FJW	Dennis RS237	Saxon Sanbec	PRL	1990	C7 Ladywood
097	G675FJW	Dennis RS237	Saxon Sanbec	PRL	1990	D8 West Bromwich
098	G676FJW	Dennis RS237	Saxon Sanbec	PRL	1990	E5 Wolverhampton
119	G677FJW	Dennis DFD133	Carmichael/Magirus DL30	TL	1989	D1 Oldbury
352	G682FJW	Freight Rover Sherpa 400	Freight Rover	PCV	1990	C3 Smethwick
353	G684FJW	Freight Rover Sherpa 400	Freight Rover	PCV	1990	Training Centre
099	H184MOK	Dennis SS237	Carmichael	PRL	1991	A6 Ward End
100	H185MOK	Dennis SS237	Carmichael	PRL	1991	B3 Coventry
101	H186MOK	Dennis SS237	Carmichael	PRL	1991	Training Centre
102	H187MOK	Dennis SS237	Carmichael	PRL	1991	C4 Bournbrook
103	H188MOK	Dennis SS237	Carmichael	PRL	1991	E1 Walsall
104	H189MOK	Dennis SS237	Carmichael	PRL	1991	E2 Bloxwich
105	H191MOK	Dennis SS237	Carmichael	PRL	1991	E8 Tettenhall
120	H204MOK	Volvo FL10	Saxon/Simon Snorkel ST300-S	ALP	1991	C4 Bournbrook
122	H207MOK	Dennis DFD237	Dennis/Magirus DL30	TL	1991	A6 Ward End
121	J43SOF	Volvo FL10	Saxon/Simon Snorkel ST300-S	ALP	1991	C7 Ladywood
250	J47SOF	Vauxhall Midi	Vauxhall	PCV	1992	B1 Solihull
251	J48SOF	Vauxhall Midi	Vauxhall	PCV	1992	D4 Halesowen
008	J56SOF	Dennis SS239	Carmichael	PRL	1992	A4 Erdington
055	J58SOF	Dennis SS239	Carmichael	PRL	1992	E7 Bilston
057	J59SOF	Dennis SS239	Carmichael	PRL	1992	B6 Binley
106	J61SOF	Dennis SS239	Carmichael	PRL	1992	C3 Smethwick
107	J62SOF	Dennis SS239	Carmichael	PRL	1992	C5 Kings Norton

The ERF EC8 range cover a wide variety of fire appliance chassis types. West Midlands operate P266EON a 6x2 example with low-line cab. This example was exhibited at the Fire 96 exhibition in Manchester. *Gavin Stewart*

108	J63SOF	Dennis SS239	Carmichael	PRL	1992	D9 Dudley
109	J64SOF	Dennis SS239	Carmichael	PRL	1992	E6 Fallings Park
001	K978XOF	Dennis Rapier TF202	Carmichael International	PRL	1992	D3
137	K979XOF	Dennis SS239	Carmichael International	PRL	1992	A1 Central Birmingham
138	K980XOF	Dennis SS239	Carmichael International	PRL	1992	A6 Ward End
139	K981XOF	Dennis SS239	Carmichael International	PRL	1992	B1 Solihull
140	K982XOF	Dennis SS239	Carmichael International	PRL	1992	B2 Sheldon
141	K983XOF	Dennis SS239	Carmichael International	PRL	1992	B5 Foleshill
142	K984XOF	Dennis SS239	Carmichael International	PRL	1992	C9 Hay Mills
143	K985XOF	Dennis SS239	Carmichael International	PRL	1992	D5 Stourbridge
144	K986XOF	Dennis SS239	Carmichael International	PRL	1992	E9 Wednesbury
235	K997XOF	Leyland DAF FA45-160	Leyland-DAF	GPV	1993	C1 Highgate
309	L961DJW	Volvo FS7.18	Multilift	PM	1994	B7 Bickenhall
310	L962DJW	Volvo FS7.18	Multilift	PM	1994	D8 West Bromwich
123	L964DJW	Volvo FL10	Saxon /Simon Snorkel ST300-S	ALP	1993	E1 Walsall
072	L971DJW	Dennis SS239	Carmichael International	PRL	1993	A7 Handsworth
073	L972DJW	Dennis SS239	Carmichael International	PRL	1993	B4 Canley
076	L973DJW	Dennis SS239	Carmichael International	PRL	1993	C4 Bournbrook
077	L974DJW	Dennis SS239	Carmichael International	PRL	1993	C6 Northfield
078	L975DJW	Dennis SS239	Carmichael International	PRL	1993	C7 Ladywood
081	L976DJW	Dennis SS239	Carmichael International	PRL	1993	D1 Oldbury
084	L977DJW	Dennis SS239	Carmichael International	PRL	1993	D2 Brierley Hill
088	L978DJW	Dennis SS239	Carmichael International	PRL	1993	D8 West Bromwich
124	M451KOV	Volvo FL10	Saxon/Simon Snorkel ST290-S	ALP	1995	E5 Wolverhampton
305	M452KOV	Volvo FS7.18	Multilift	PM	1995	C9 Hay Mills
306	M453KOV	Volvo FS7.18	Multilift	PM	1995	E3 Willenhall
035	M464KOV	Dennis SS239	John Dennis	PRL	1995	B1 Solihull
036	M465KOV	Dennis SS239	John Dennis	PRL	1995	B3 Coventry
037	M466KOV	Dennis SS239	John Dennis	PRL	1995	B7 Bickenhall
038	M467KOV	Dennis SS239	John Dennis	PRL	1995	D4 Halesowen
039	M468KOV	Dennis SS239	John Dennis	PRL	1995	D9 Dudley
040	M469KOV	Dennis SS239	John Dennis	PRL	1995	C1 Highgate
041	M470KOV	Dennis SS239	John Dennis	PRL	1995	E1 Walsall
042	M471KOV	Dennis SS239	John Dennis	PRL	1995	E5 Wolverhampton
200	M472KOV	Dennis Dart 9.8SDL3017	Plaxton Pointer	PCV	1995	Training Centre

West Midlands Fire Service are the first British brigade to purchase the Dennis Sabre XL model which features an enlarged rear crew cab than the standard model. The vehicle is shown when it made its debut in July 1996 at the Cannon Hill public muster day. N187SDA is based at Tipton. *Bill Murray*

201	M473KOV	LDV 400	LDV 15-seat	PCV	1995	A1 Central Birmingham
202	M474KOV	LDV 400	LDV 15-seat	PCV	1995	E1 Walsall
203	N178SDA	LDV 400	LDV 15-seat	PCV	1996	Training Centre
020	N182SDA	Dennis Sabre XL	John Dennis	PRL	1996	A1 Central Birmingham
021	N183SDA	Dennis Sabre XL	John Dennis	PRL	1996	A2 Aston
022	N184SDA	Dennis Sabre XL	John Dennis	PRL	1996	B6 Binley
023	N185SDA	Dennis Sabre XL	John Dennis	PRL	1996	C6 Northfield
026	N186SDA	Dennis Sabre XL	John Dennis	PRL	1996	D6 Cradeley Heath
027	N187SDA	Dennis Sabre XL	John Dennis	PRL	1996	D7 Tipton
029	N188SDA	Dennis Sabre XL	John Dennis	PRL	1996	E3 Willenhall
030	N189SDA	Dennis Sabre XL	John Dennis	PRL	1996	E4 Aldridge
110	N192SDA	Volvo FL10	Saxon/Simon Snorkel ST290-S	ALP	1996	D2 Brierley Hill
301	P264EON	ERF EC8	Multilift	PM	1996	
302	P265EON	ERF EC8	Multilift	PM	1996	
112	P266EON	ERF EC8	Saxon/Simon Snorkel ST340-S	APL	1996	
125	P267EON	ERF EC8	Saxon/Simon Snorkel ST340-S	APL	1996	
118	P268EON	ERF EC8	Saxon/Simon Snorkel ST340-S	APL	1996	
044	P269EON	Dennis Sabre XL 4.2m	John Dennis	PRL	1996	
045	P270EON	Dennis Sabre XL 4.2m	John Dennis	PRL	1996	
046	P271EON	Dennis Sabre XL 4.2m	John Dennis	PRL	1996	
048	P272EON	Dennis Sabre XL 4.2m	John Dennis	PRL	1996	
049	P273EON	Dennis Sabre XL 4.2m	John Dennis	PRL	1996	
051	P274EON	Dennis Sabre XL 4.2m	John Dennis	PRL	1996	
052	P275EON	Dennis Sabre XL 4.2m	John Dennis	PRL	1996	
053	P276EON	Dennis Sabre XL 4.2m	John Dennis	PRL	1996	
054	P277EON	Dennis Sabre XL 4.2m	John Dennis	PRL	1996	

Previous Registrations:

Q68VOE	POB984R	Q69VOE	POB983R	Q70VOE	UOF638S

085/6/9 were originally CU/ET/BAT combined appliances. 234/6 were both originally buses with West Midlands PTE

West Sussex Fire Brigade continue to procure batches of Dennis Rapier Water Tender Ladders fron John Dennis coachbuilders. The latest batch wear the new livery that is also carried by the new BA Support Units, and the Incident Control Unit. A00045 (N139DPX) is based at Crawley. *Gavin Stewart*

The generally lower floor line, and easy accessibility onto buses can be advantageous in certain fire fighting applications. One example is this Incident Command Unit recently delivered to West Sussex FB. A 9 metre Dennis Dart bus chassis is combined with coachwork by Leicester Carriage Company on A00150 (N143DPX). It was photographed prior to going on the run at Bognor Regis. *Simon Pearce*

WEST SUSSEX FIRE BRIGADE

West Sussex County Fire Brigade, Northgate, Chichester PO19 1BD

Arundel Castle	Shand Mason No1	Steam Pump	P	1893	Preserved at A01 Worthing
A00002 H878BOT	Dennis SS293	Reynolds Boughton	WrL	1991	A02 East Preston
A00003 H879BOT	Dennis SS293	Reynolds Boughton	WrL	1991	A17 Chichester
A00004 H880BOT	Dennis SS293	Reynolds Boughton	WrL	1991	A07 Littlehampton
A00005 TPO514M	Range Rover 6x4	Carmichael Commando	RT	1974	B12 Horsham
A00008 G265VBP	Mercedes-Benz 917AF	Mountain Range	CT	1990	B08 Crawley
A00009 H881BOT	Dennis SS293	Reynolds Boughton	WrL	1991	B12 Horsham
A00013 OTR213S	Dennis R130	Dennis	WrL	1978	Driving School
A00015 TCR349T	Dennis R133	Dennis	WrL	1979	A Reserve
A00017 CBP967W	Dennis RS133	Dennis	WrL	1981	A Reserve
A00018 CBP968W	Dennis RS133	Dennis	WrL	1981	A03 Lancing
A00019 E611JBP	Dennis SS137	Carmichael	WrL	1988	B08 Crawley
A00020 E612JBP	Dennis SS137	Carmichael	WrL	1988	B10 Horley
A00021 E613JBP	Dennis SS137	Carmichael	WrL	1988	A01 Worthing
A00022 E614JBP	Dennis SS137	Carmichael	WrL	1988	A20 Bognor Regis
A00023 F21OCR	Dennis DS153	HCB-Angus	WrL	1989	B13 Billinghurst
A00024 F22OCR	Dennis DS153	HCB-Angus	WrL	1989	D28 Hurstpierpoint
A00025 F23OCR	Dennis DS153	HCB-Angus	WrL	1989	A06 Arundel
A00026 F24OCR	Dennis DS153	HCB-Angus	WrL	1989	C19 Midhurst
A00027 G261VBP	Dennis SS137	Reynolds Boughton	WrL	1990	A03 Lancing
A00028 G262VBP	Dennis SS137	Reynolds Boughton	WrL	1990	D23 East Grinstead
A00029 G263VBP	Dennis SS137	Reynolds Boughton	WrL	1990	A05 Shoreham by Sea
A00030 M469XCR	Dennis Rapier TF202	John Dennis	WrL	1994	A21 Petworth
A00031 G264VBP	Dennis SS137	Reynolds Boughton	WrL	1990	D24 Haywards Heath
A00032 J32GRV	Dennis Rapier TF202	John Dennis	WrL	1992	B08 Crawley
A00033 J74GRV	Dennis Rapier TF202	John Dennis	WrL	1992	A01 Worthing
A00035 J75GRV	Dennis Rapier TF202	John Dennis	WrL	1992	D25 Burgess Hill
A00036 J76GRV	Dennis Rapier TF202	John Dennis	WrL	1992	D27 Keymer
A00037 L83PTR	Dennis Rapier TF202	John Dennis	WrL	1994	C20 Bognor Regis
A00038 L84PTR	Dennis Rapier TF202	John Dennis	WrL	1994	B10 Horley
A00039 L85PTR	Dennis Rapier TF202	John Dennis	WrL	1994	A04 Findon
A00040 L86PTR	Dennis Rapier TF202	John Dennis	WrL	1994	D22 Turners Hill
A00041 M470XCR	Dennis Rapier TF202	John Dennis	WrL	1994	B14 Storrington
A00042 M471XCR	Dennis Rapier TF202	John Dennis	WrL	1994	A05 Shoreham by Sea
A00043 M472XCR	Dennis Rapier TF202	John Dennis	WrL	1994	A19 Midhurst
A00044 N138DPX	Dennis Rapier R411	John Dennis	WrL	1995	D24 Haywards Heath
A00045 N139DPX	Dennis Rapier R411	John Dennis	WrL	1995	B08 Crawley
A00046 N140DPX	Dennis Rapier R411	John Dennis	WrL	1995	D23 East Grinstead
A00047 N141DPX	Dennis Rapier R411	John Dennis	WrL	1995	A01 Worthing
A00048 TCR351T	Dennis R133	Dennis	WrL	1979	Training Centre
A00052 XOT158V	Dennis RS133	Dennis	WrL	1980	Driving School
A00053 XOT159V	Dennis RS133	Dennis	WrL	1980	A Reserve
A00054 XOT160V	Dennis RS133	Dennis	WrL	1980	Driving School
A00055 CBP969W	Dennis RS133	Dennis	WrL	1981	B Reserve
A00056 H197GKM	Iveco-Ford Cargo 1721	John Dennis/Reynolds Boughton	WrC	1991	A07 Littlehampton
A00057 FPO24X	Dennis DF133	Dennis	FoT	1981	A05 Shoreham by Sea
A00058 FPO14X	Bedford TM/USG 4x4	West Sussex FB	HL	1982	B10 Horley
A00059 KCR586Y	Dennis RS133	Dennis	WrL	1983	A07 Littlehampton
A00060 KCR587Y	Dennis RS133	Dennis	WrL	1983	D25 Burgess Hill
A00061 KCR588Y	Dennis RS133	Dennis	WrL	1983	B Reserve
A00062 F25OCR	Dennis F127	Saxon/Simon Snorkel SS263	HP	1989	A01 Worthing
A00063 TCR354T	Dennis Delta 2	Dennis/Simon Snorkel SS220	HP	1979	C17 Chichester
A00064 FPO23X	Dennis Delta 2	Dennis/Simon Snorkel SS220	HP	1981	B12 Horsham
A00066 B416UBP	Dennis RS133	Dennis	WrL	1984	B Reserve
A00067 B417UBP	Dennis RS133	Dennis	WrL	1984	A02 East Preston
A00068 B418UBP	Dennis RS133	Dennis	WrL	1984	A17 Chichester
A00069 B419UBP	Dennis RS133	Dennis	WrL	1984	B12 Horsham
A00070 C351XRV	Dennis DS151	Carmichael	WrL	1986	B09 Partridge Green
A00071 C352XRV	Dennis DS151	Carmichael	WrL	1986	B14 Storrington
A00072 C353XRV	Dennis DS151	Carmichael	WrL	1986	C15 Selsey
A00073 C354XRV	Dennis DS151	Carmichael	WrL	1986	C16 East Wittering

A00074	D554DOR	Dennis DS151	Carmichael	WrL	1987	A21 Petworth
A00075	D555DOR	Dennis DS151	Carmichael	WrL	1987	B11 Steyning
A00076	D556DOR	Dennis DS151	Carmichael	WrL	1987	C18 Bosham
A00077	D557DOR	Dennis DS151	Carmichael	WrL	1987	B26 Henfield
A00078	K658MBP	Mercedes-Benz 1124AF	John Dennis/HIAB	HRT	1993	C17 Chichester
A00079	K659MBP	Mercedes-Benz 1124AF	John Dennis/HIAB	HRT	1993	A01 Worthing
A00150	N143DPX	Dennis Dart 9SDL3056	Leicester Carriage Co	ICU	1996	A20 Bognor Regis
A00151	P	Dennis Dart 9SDL3056	Leicester Carriage Co	ICU	1996	
A00152	M477XCR	Iveco-Ford Super Cargo 170E23	Leicester Carriage Co	BASU	1995	B12 Horsham
A00153	M478XCR	Iveco-Ford Super Cargo 170E23	Leicester Carriage Co	BASU	1995	A17 Chichester
A00154	N142DPX	Iveco-Ford Super Cargo 170E23	Leicester Carriage Co	BASU	1996	A05 Shoreham by Sea
A00180	TCR355T	Land Rover Series III 109	West Sussex FB	L4T	1979	B12 Horsham
A00181	TCR356T	Land Rover Series III 109	West Sussex FB	L4T	1979	D23 East Grinstead
A00182	TCR357T	Land Rover Series III 109	West Sussex FB	L4T	1979	A20 Bognor Regis
A00183	TCR358T	Land Rover Series III 109	West Sussex FB	L4T	1979	B08 Crawley
A00184	XOT141V	Land Rover Series III 109	West Sussex FB	L4T	1981	B14 Storrington
A00185	XOT142V	Land Rover Series III 109	West Sussex FB	L4T	1981	A21 Petworth
A00186	XOT144V	Land Rover Series III 109	West Sussex FB	L4T	1981	C15 Selsey
A00187	XOT145V	Land Rover Series III 109	West Sussex FB	L4T	1981	A05 Shoreham by Sea
A00188	XOT146V	Land Rover Series III 109	West Sussex FB	L4P	1981	A19 Midhurst
A00189	XOT147V	Land Rover Series III 109	West Sussex FB	L4T	1981	D24 Haywards Heath
A00190	F326PPG	Land Rover 110	West Sussex FB (1995)	CSU	1989	C16 East Wittering
A00191	FPO22X	Land Rover Series III 109	West Sussex FB	CSU	1982	Headquarters
A00245	M476XCR	Ford Transit	Ford Minibus	PCV	1994	A01 Worthing
A00252	H874BOT	Ford Transit	Ford Minibus	PCV	1990	B12 Horsham
A00313	OTR215S	Ford A0609	Hawson Garner/West Sussex FB	CU	1977	A20 Bognor Regis
A00332	G266VBP	Iveco-Ford Cargo 0813	Hillbrow/Ratcliff	GPV	1990	B10 Horley
A00333	G267VBP	Iveco-Ford Cargo 0813	Hillbrow/Ratcliff	GPV	1990	A20 Bognor Regis
A00337	XOT151V	Ford A0609	Hawson Garner/West Sussex FB	CU	1980	D24 Haywards Heath
A		Mercedes-Benz 1124AF	John Dennis	HRT	1996	
A		Mercedes-Benz 1124AF		OSU	1996	
A		Mercedes-Benz 1124AF		HL	1996	
A		Moffat Mounty All-Terrain M....		FLT	1996	
A		Moffat Mounty All-Terrain M....		FLT	1996	
A	Boat	Moffat Mounty All-Terrain M....	Boat and Trailer	IRBt	19	

Two new Breathing Apparatus Support Units were placed on the run with West Sussex in 1995. These have replaced the Ford D series BA Control Vans. A00153 (M478XCR) from Chichester fire station has an Iveco-Ford Super Cargo chassis and Leicester Carriage Company bodywork. Note the addition of yellow and black chequers on this vehicle.
Jef Johnson

The Fire Brigade Handbook

WEST YORKSHIRE FIRE SERVICE

West Yorkshire County Fire Brigade, Oakroyd Hall, Birkenshaw BD11 2DY

007	C859GCX	Dodge G16C	Carmichael/Magirus DL30	TL	1986	D Reserve
008	C860GCX	Dodge G16C	Carmichael/Magirus DL30	TL	1986	C Reserve
	C486KHD	Bedford TL	Customline	FPU	1986	Fire Prevention
004	D89SCX	Renault-Dodge G16C	Saxon/Simon Snorkel SS263	HP	1986	A20 Leeds Central
	D93SCX	Bedford TL	HCB-Angus	WrL	1987	Driving School
043	D94SCX	Bedford TL	HCB-Angus	WrT	1987	B45 Ilkley
	D96SCX	Bedford TL	HCB-Angus	WrL	1987	Training Centre
	D97SCX	Bedford TL	HCB-Angus	WrL	1987	Training Centre
044	D98SCX	Bedford TL	West Yorkshire FS	WrL	1987	Training Centre
112	D155SCX	Bedford TL750	HCB-Angus	ET	1987	D Reserve
009	E159DVH	Volvo FL6.16	Angloco/Metz DL30	TL	1988	A21 Bramley, Leeds
047	E164DVH	Volvo FL6.14	West Yorkshire FS	WrT	1988	B52 Silsden
059	E165DVH	Volvo FL6.14	HCB-Angus	WrL	1987	C Reserve
030	E166DVH	Volvo FL6.14	HCB-Angus	WrL	1988	B43 Harworth
096	E167DVH	Volvo FL6.14	HCB-Angus	WrL	1988	D87 Mirfield
048	E168DVH	Volvo FL6.14	HCB-Angus	WrL	1988	C69 Meltham
084	E169DVH	Volvo FL6.14	HCB-Angus	WrL	1988	D84 Featherstone
091	E170DVH	Volvo FL6.14	HCB-Angus	WrL	1988	D Reserve
040	E171DVH	Volvo FL6.14	HCB-Angus	WrL	1988	B48 Otley
	E172DVH	Volvo FL4	Devcoplan	FPU	1988	Fire Prevention
113	E174DVH	Bedford TL850	Devcoplan	ET	1988	A Reserve

Over 90 Volvo appliances have transformed the West Yorkshire Fire Service fleet since 1988. The most common combination are Water Tender Ladders with bodywork built by HCB-Angus as illustrated by H219UCX. In addition to electronic sirens, many of West Yorkshires newer appliances also feature a very effective air horn. *Graham Hopwood*

One of the appliances bought to meet the 1995 requirement for West Yorkshire pump appliances is 138, N979FWU, a Volvo FL6 with Emergency One bodywork. The vehicle carries yellow and blue reflective stripes and is based at Illingworth fire station. *Tim Ansell*

The first Volvo-based aerial appliance with West Yorkshire was 009, E159DVH, a FL6 with Angloco bodywork and Metz turntable ladder. It is now based in Leeds after several years at Wakefield. *Tim Ansell*

A water relay is a long run of hose used to augment insufficient water supplies at a fireground. A development of the hose layer is the Hydro-Subsystem used by West Yorkshire Fire Service. It features half-width pod units for hose laying or pumping. This view shows 120 (K593SJX) a Volvo FL6.18 PM with combi-pod 126 and half pods 061 and 123 at Bradford Central. *Karl Sillitoe*

042	E176DVH	Volvo FL6.14	HCB-Angus	WrL	1988	B45 Ilkley
102	E177DVH	Volvo FL6.14	HCB-Angus	WrL	1988	D Reserve
050	E178DVH	Volvo FL6.14	HCB-Angus	WrL	1988	B52 Silsden
105	E179DVH	Volvo FL6.14	HCB-Angus	WrL	1988	D Reserve
028	E180DVH	Volvo FL6.14	HCB-Angus	WrL	1988	Driving School
093	E181DVH	Volvo FL6.14	HCB-Angus	WrL	1988	D85 Hemsworth
019	E182DVH	Volvo FL6.14	HCB-Angus	WrL	1988	A Reserve
003	F847SHD	Volvo FL6.17	Saxon/Simon Snorkel SS263	HP	1988	A20 Leeds Central
	F849SHD	Volvo FL6.14	West Yorkshire FS	DTV	1989	Driving School
041	F850SHD	Volvo FL6.14	West Yorkshire FS	WrT	1989	B46 Keighley
114	F855SHD	Renault-Dodge G08	Devcoplan	ET	1989	D80 Wakefield
089	F856SHD	Volvo FL6.14	HCB-Angus	WrT	1988	D83 Dewsbury
085	F857SHD	Volvo FL6.14	HCB-Angus	WrL	1989	D81 Batley
022	F858SHD	Volvo FL6.14	HCB-Angus	WrL	1989	A Reserve
074	F859SHD	Volvo FL6.14	HCB-Angus	WrL	1989	C70 Mytholmroyd
087	F860SHD	Volvo FL6.14	HCB-Angus	WrL	1989	D82 Castleford
071	F861SHD	Volvo FL6.14	HCB-Angus	WrL	1989	C Reserve
025	F862SHD	Volvo FL6.14	HCB-Angus	WrL	1989	A27 Morley
078	F863SHD	Volvo FL6.14	HCB-Angus	WrL	1989	C Reserve
017	F864SHD	Volvo FL6.14	HCB-Angus	WrL	1989	C65 Hebden Bridge
024	F865SHD	Volvo FL6.14	HCB-Angus	WrL	1989	A25 Hunslet, Leeds
034	F866SHD	Volvo FL6.14	HCB-Angus	WrL	1989	Reserve
080	F867SHD	Volvo FL6.14	HCB-Angus	WrL	1989	C74 Todmorden
010	F868SHD	Volvo FL6.17	Angloco/Metz DL30	TL	1989	D81 Batley
027	G340FCP	Volvo FL6.14	HCB-Angus	WrT	1989	A26 Moortown, Leeds
049	G341FCP	Volvo FL6.14	HCB-Angus	WrL	1989	B49 Pudsey
083	G342FCP	Volvo FL6.14	HCB-Angus	WrT	1989	D80 Wakefield
031	G343FCP	Volvo FL6.14	HCB-Angus	WrL	1989	A29 Weatherby
094	G344FCP	Volvo FL6.14	HCB-Angus	WrL	1989	D86 Knottingley
046	G345FCP	Volvo FL6.14	West Yorkshire FS	WrT	1989	B47 Odsal, Bradford
062	G346FCP	Volvo FL6.14	West Yorkshire FS	WrL	1990	C62 Cleckheaton
115	G350FCP	Volvo FL6.8	Devcoplan	ET	1990	A24 Gipton, Leeds
116	G351FCP	Volvo FL6.8	Devcoplan	ET	1990	C60 Huddersfield
018	G371FCP	Volvo FL6.14	HCB-Angus	WrL	1990	A22 Cookridge, Leeds
016	G372FCP	Volvo FL6.14	HCB-Angus	WrT	1990	A21 Bramley, Leeds
054	G373FCP	Volvo FL6.14	HCB-Angus	WrL	1990	B51 Shipley
023	G374FCP	Volvo FL6.14	HCB-Angus	WrL	1990	A25 Hunslet, Leeds
065	G375FCP	Volvo FL6.14	HCB-Angus	WrL	1990	C64 Halifax
033	G376FCP	Volvo FL6.14	HCB-Angus	WrL	1990	B Reserve
053	G377FCP	Volvo FL6.14	HCB-Angus	WrL	1990	B51 Shipley
029	G378FCP	Volvo FL6.14	HCB-Angus	WrL	1990	A28 Stanks Drive, Leeds
058	G379FCP	Volvo FL6.14	HCB-Angus	WrL	1990	C60 Huddersfield
117	H178UCX	Volvo FL6.14	West Yorkshire FS	HRU	1990	B44 Idle, Bradford
011	H179UCX	Volvo FL6.17	Angloco/Metz DL30	TL	1990	B46 Keighley
012	H180UCX	Volvo FL6.17	Angloco/Metz DL30	TL	1990	C64 Halifax
020	H181UCX	Volvo FL6.14	HCB-Angus	WrL	1990	A20 Leeds Central
090	H182UCX	Volvo FL6.14	HCB-Angus	WrL	1990	D83 Dewsbury
086	H183UCX	Volvo FL6.14	HCB-Angus	WrL	1990	D81 Batley
039	H184UCX	Volvo FL6.14	West Yorkshire FS	WrL	1991	B46 Keighley
	H185UCX	Volvo FL6.14	West Yorkshire FS	DTV	1991	Driving School
103	H217UCX	Volvo FL6.14	HCB-Angus	WrL	1991	D92 South Emsall
063	H218UCX	Volvo FL6.14	HCB-Angus	WrL	1991	C63 Elland
082	H219UCX	Volvo FL6.14	HCB-Angus	WrL	1991	D80 Wakefield
064	H220UCX	Volvo FL6.14	HCB-Angus	WrT	1991	C64 Halifax
013	H221UCX	Volvo FL6.14	HCB-Angus	WrT	1991	A20 Leeds Central
021	J784GJX	Volvo FL6.14	HCB-Angus	WrL	1991	C71 Skelmanthorpe
026	J785GJX	Volvo FL6.14	HCB-Angus	WrL	1991	A26 Moortown, Leeds
057	J786GJX	Volvo FL6.14	HCB-Angus	WrT	1991	C60 Huddersfield
070	J787GJX	Volvo FL6.14	HCB-Angus	WrL	1991	D88 Normanton
098	J788GJX	Volvo FL6.14	West Yorkshire FS	WrL	1991	D89 Ossett
038	J789GJX	Volvo FL6.14	West Yorkshire FS	WrL	1991	B44 Idle, Bradford
118	J791GJX	Volvo FL6.14	Powell Duffryn Multilift	PM	1992	D83 Dewsbury
119	WY1	Volvo FL6.14	Powell Duffryn Multilift	PM	1992	C62 Cleckheaton
051	J911MJX	Volvo FL6.14	HCB-Angus	WrL	1992	B50 Rawdon
100	J914MJX	Volvo FL6.14	HCB-Angus	WrL	1992	D90 Pontefract
002	J915MJX	Volvo FL6.18	Bedwas/Simon Snorkel SS263	HP	1992	B40 Bradford Central
001	J916MJX	Volvo FL6.18	Bedwas/Simon Snorkel SS263	HP	1992	C60 Huddersfield
076	K585SJX	Volvo FL6.14	HCB-Angus	WrL	1992	C72 Slaithwaite
101	K586SJX	Volvo FL6.14	HCB-Angus	WrL	1992	D91 Rothwell
015	K587SJX	Volvo FL6.14	HCB-Angus	WrL	1992	A21 Bramley, Leeds

032	K588SJX	Volvo FL6.14	HCB-Angus	WrL	1992	B40 Bradford Central
014	K589SJX	Volvo FL6.14	HCB-Angus	WrL	1992	A23 Garforth
060	K590SJX	Volvo FL6.14	HCB-Angus	WrL	1992	C61 Brighouse
122	K591SJX	Mercedes-Benz Unimog 2150L	West Yorkshire FS/Atlas 100.1	SIU	1992	C63 Elland
121	K592SJX	Volvo FL6.18	Powell Duffryn Multilift Machook	PM	1992	D Reserve
120	K593SJX	Volvo FL6.18	Powell Duffryn Multilift Machook	PM	1992	B40 Bradford Central
035	K594SJX	Volvo FL6.14	West Yorkshire FS	WrL	1993	B41 Bingley
036	K572UCX	Volvo FL6.14	West Yorkshire FS	WrL	1993	B42 Fairweather Gn, Bradford
081	K575UCX	Volvo FL6.14	West Yorkshire FS	WrL	1993	B47 Odsal, Bradford
097	L728EJX	Volvo FL6.18	Powell Duffryn Multilift Machook	PM	1992	A25 Hurslet, Leeds
099	L729EJX	Volvo FL6.18	Powell Duffryn Multilift Machook	PM	1992	Driving School/Reserve
104	L730EJX	Volvo FL6.14	HCB-Angus	WrL	1993	A24 Gipton, Leeds
	Un-registered	Case Skid-steer loader	Case	SSL	1993	A26 Moortown, Leeds
136	N977FWU	Volvo FL6.14	Emergency One	WrL	1995	B40 Bradford Central
137	N978FWU	Volvo FL6.14	Emergency One	WrT	1995	A24 Gipton, Leeds
138	N979FWU	Volvo FL6.14	Emergency One	WrL	1995	C67 Illingworth, Halifax
	P653PWW	Volvo FL6.14	Emergency One	WrL	1995	D83 Dewsbury
	P654PWW	Volvo FL6.14	Emergency One	WrL	1995	D82 Castleford
	P655PWW	Volvo FL6.14	Emergency One	WrL	1995	B45 Keighley

Pods and containers:

052	Pod	Locomotors	Incident Support Unit	ISU	1992	A25 Hunslet, Leeds
125	Pod	Kryton	Hydro-Sub Combi-Pod	CP	1993	A25 Hunslet, Leeds
128	Half Pod	West Yorkshire FS	Hydro-Sub Hose Layer (1Km)	HL	1993	A25 Hunslet, Leeds
130	Half Pod	West Yorkshire FS	Hydro-Sub Hose Layer (1Km)	HL	1993	A25 Hunslet, Leeds
061	Half Pod	West Yorkshire FS	Hose Layer (2Km)	HL	1996	B40 Bradford Central
088	Pod	Bill Scott	Personnel Carrier/Refreshment	PCR	1993	B40 Bradford Central
123	Half Pod	West Yorkshire FS	Hydro-Sub Pumping Unit	HSU	1993	B40 Bradford Central
126	Pod	Kryton	Hydro-Sub Combi-Pod	CP	1993	B40 Bradford Central
005	Pod	West Yorkshire FS	Major Rescue Unit	MRU	1995	C62 Cleckheaton
037	Pod	Leicester Carriage Co	Control Unit	CU	1992	C62 Cleckheaton
095	Pod	Devcoplan	Foram Carrier	FoT	1993	C62 Cleckheaton
131	Container	West Yorkshire FS	Sewer Rescue	SR	1992	C63 Elland
132	Container	West Yorkshire FS	Moorland/Farm Incident	MF	1992	C63 Elland
133	Container	West Yorkshire FS	Moorland/Farm Incident	MF	1992	C63 Elland
134	Container	West Yorkshire FS	General Rescue	GR	1992	C63 Elland
135	Container	West Yorkshire FS	Hose Layer	HL	1992	C63 Elland
045	Pod	Leicester Carriage Co	Salvage Unit	ST	1993	D83 Dewsbury
055	Pod	West Yorkshire FS	Flat Bed Lorry	FBL	1993	D83 Dewsbury
066	Pod	Locomotors	Hazardous Materials Unit	HMU	1992	D83 Dewsbury
124	Pod	West Yorkshire FS	Hydro-Sub Hose Layer (2Km)	HL	1993	D83 Dewsbury
006	Pod	West Yorkshire FS	Foam Tender	FoT	1995	D91 Rothwell
106	Pod	West Yorkshire FS	BA Control Unit	BACU	1994	D91 Rothwell
056	Pod	Lynton Commercials	Occupational Health Unit	OHU	1994	Headquarters
111	Pod	Bill Scott/West Yorkshire FS	BA Training Unit	BA(T)	1993	Headquarters
129	Pod	Bill Scott/West Yorkshire FS	Control Unit	CU(T)	1994	Headquarters

Notes:
Combi-pod 125 carries two pods (128 & 130); Combi-pod 126 carries one pod (123), Combi-pods 125 & 126 are carried by two Prime Movers; Up to 3 containers (131-135) are carried as required by Unimog SIU appliance, alternatively it can carry any Half-pod, or the Skid Steer loader normally carried upon Pod 045.

The Fire Brigade Handbook

WILTSHIRE FIRE BRIGADE

Manor House, Potterne, Devizes SN10 5PP

HMR765	Dennis F12	Dennis	PE	1951	Preserved
JHR486E	Land Rover Series IIA 109	Land Rover	L4V	1967	4/5 Warminster
DWV26L	Land Rover Series III 109	Land Rover	L4V	1972	3/4 Mere
PMR713M	Dodge K850	Merryweather/Wiltshire FB	DCU	1974	1/1 Swindon
GHR618N	Dodge K850	Merryweather/Wiltshire FB	DCU	1974	4/4 Westbury
JMW783P	Land Rover Series III 109	Land Rover	L4V	1976	1/2 Cricklade
MMR600R	Dodge K850	Ray Smith	PM	1977	Headquarters
OHR621R	Dodge K1113	Cheshire Fire Engineering	WrL	1977	Reserve
UMW326T	Dodge K1113	Cheshire Fire Engineering	WrL	1978	3/6 Ludgershall
UMW329T	Dodge K1113	Cheshire Fire Engineering	WrL	1978	Training Centre
XMW758T	Dodge G1690	Carmichael/Magirus DL30	TL	1979	Reserve
JPH84V	Leyland Clydesdale	Wincanton Tankers	WrC	1979	1/6 Wooton Bassett
YMW61V	Dodge G13C	HCB-Angus	WrL	1979	1/4 Stratton St Margarets
YMW62V	Dodge G1313	HCB-Angus	WrL	1979	Reserve
YMW63V	Dodge G1313	HCB-Angus	WrL	1979	Reserve
YMW64V	Dodge G1313	HCB-Angus	WrL	1979	Reserve
FHR622W	Dodge S56	Carmichael	RT	1980	Reserve
GMW180W	Dodge G1313	Cheshire Fire Engineering	WrL	1980	1/6 Wooton Bassett
GMW181W	Dodge G1313	Cheshire Fire Engineering	WrL	1980	4/4 Westbury
GMW182W	Dodge G1313	Cheshire Fire Engineering	WrL	1980	3/4 Mere
WYC983W	Leyland Clydesdale	Wincanton Tankers	WrC	1980	4/5 Warminster
XYA369W	Leyland Clydesdale	Wincanton Tankers	WrC	1980	3/2 Wilton
MHR849X	Dodge G1313	HCB-Angus	WrL	1982	Reserve
NAM362X	Dodge G09	Benson	CU/CAV	1982	2/1 Chippenham
RMW21Y	Dennis SS133	Dennis	WrL	1983	3/2 Wilton
RMW22Y	Dennis SS133	Dennis	WrL	1983	Training Centre
RMW23Y	Dennis SS133	Dennis	WrL	1983	
A512EPB	Leyland Clydesdale	Wincanton Tankers(1992)	WrC	1984	1/3 Ramsbury
A352SFB	Land Rover 110	Land Rover	L4V	1984	4/6 Devizes
A955XOB	Dennis Delta DF1616	Multilift	PM	1984	Training Centre
B757DMR	Dodge G13C	HCB-Angus	WrL	1985	4/2 Bradford-upon-Avon
B758DMR	Dodge G13C	HCB-Angus	WrL	1985	2/4 Calne
B759DMR	Dodge G13C	HCB-Angus	WrL	1985	4/5 Warminster

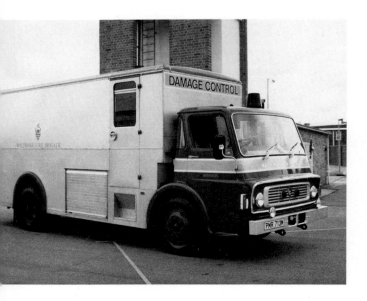

Wiltshire's Damage Control Units were converted from Emergency Tenders. Both machines have Dodge K850 chassis and Merryweather bodywork. The K series chassis, and Merryweather coachwork are becoming rare among UK brigades. PMR713M is based at Swindon while its sister appliance is at Westbury.
Simon Pearce

B622VHY	Land Rover 110	Land Rover	L4V	1985	3/2 Wilton
C107HTU	Land Rover 110	Land Rover	L4V	1985	4/1 Trowbridge
C108HTU	Land Rover 110	Land Rover	L4V	1986	1/3 Ramsbury
C96KMR	Dodge G13C	HCB-Angus	WrL	1986	2/3 Malmesbury
C97KMR	Dodge G13C	HCB-Angus	WrL	1986	3/6 Ludgershall
C98KMR	Dodge G13C	HCB-Angus	WrL	1986	2/2 Corsham
D307PAM	Renault-Dodge G13C	HCB-Angus	WrL	1987	1/5 Marlborough
D308PAM	Renault-Dodge G13C	HCB-Angus	WrL	1987	1/1 Swindon
D309PAM	Renault-Dodge G13C	HCB-Angus	WrL	1987	4/6 Devizes
D686RMR	Renault-Dodge G13C	HCB-Angus	WrL	1987	3/7 Pewsey
D687RMR	Land Rover 110	Land Rover	L4V	1986	2/1 Chippenham
D689RMR	Land Rover 110	Land Rover	L4V	1987	1/7 Westlea, Swindon
D690RMR	Land Rover 110	Land Rover	L4V	1987	3/5 Amesbury
E553XMR	Renault-Dodge G13TC	Mountain Range	WrL	1988	1/2 Cricklade
E554XMR	Renault-Dodge G13TC	Mountain Range	WrL	1988	4/3 Melksham
E555XMR	Renault-Dodge G13TC	Locomotors Multilift	PM	1988	3/1 Salisbury
Pod	Ray Smith/Mountain Range	Exhibition Unit	ExU	1988	Headquarters
Pod	Multilift	Flat-bed Lorry	FBL	1988	1/1 Swindon
Pod	Multilift	Flat-bed Lorry	FBL	1988	3/1 Salisbury
Pod	Devcoplan	Foam Tender	FoT	1988	3/1 Salisbury
Pod	Locomotors	Incident Control Unit	ICU	1988	3/1 Salisbury
G459NMW	Mercedes-Benz 2228	Angloco/Bronto Skylift 28-2Ti	AP	1990	3/1 Salisbury
G462NMW	Mercedes-Benz 920AF	Mountain Range	CP	1990	4/1 Trowbridge
G463NMW	Mercedes-Benz 917AF	Mountain Range	CP	1990	4/5 Warminster
G464NMW	Mercedes-Benz 917AF	Mountain Range	RT	1990	3/1 Salisbury
G465NMW	Mercedes-Benz 1120AF	Mountain Range	WrL	1990	2/2 Corsham
G466NMW	Mercedes-Benz 920AF	Mountain Range	CP	1990	3/1 Salisbury
G467NMW	Mercedes-Benz 920AF	Mountain Range	CP	1990	2/1 Chippenham
G470NMW	Mercedes-Benz 920AF	Rosenbauer	RT	1990	2/1 Chippenham
G471NMW	Mercedes-Benz 917AF	Mountain Range	RT	1990	1/7 Westlea, Swindon
G522UEU	Peugeot-Talbot Express	Wiltshire FB (1995)	CIU	1990	1/4 Stratton St Margarets
H138WMW	Dennis SS237	Mountain Range	WrL	1990	1/3 Ramsbury
J288HMR	Mercedes-Benz 1124AF	Reliance Mercury	WrL	1992	3/5 Amesbury
J289HMR	Mercedes-Benz 1124AF	Reliance Mercury	WrL	1992	4/6 Devizes
J290HMR	Mercedes-Benz 1124AF	Reliance Mercury	WrL	1992	1/5 Marlborough
J291HMR	Land Rover Defender 110 Tdi	Land Rover	L4V	1992	Headquarters
J292HMR	Land Rover Defender 110 Tdi	Land Rover	L4V	1992	3/1 Salisbury
J913JMR	Land Rover Defender 110 Tdi	Land Rover	L4V	1992	2/3 Malmesbury
J915HMR	Mercedes-Benz 1124AF	Halton HFE	WrL	1992	4/1 Trowbridge
J916HMR	Mercedes-Benz 1124AF	Halton HFE	WrL	1992	3/3 Tisbury
K627OAM	Mercedes-Benz 1124AF	Halton HFE	WrL	1993	1/4 Stratton St Margaret
K628OAM	Mercedes-Benz 1124AF	Halton HFE	WrL	1993	1/1 Swindon
K629OAM	Scania P93ML-280R 6x4	Saxon Sanbec/Italmec NCS 29m	PHP	1993	1/1 Swindon
L25TMW	Mercedes-Benz 1124AF	Devcoplan	RT	1993	4/1 Trowbridge
L850WMR	Mercedes-Benz 1124F	Emergency One	WrL/R	1994	3/5 Amesbury
L851WMR	Mercedes-Benz 1124F	Emergency One	WrL/R	1994	1/7 Westlea, Swindon
L852WMR	Mercedes-Benz 1124F	Emergency One	WrL/R	1994	3/1 Salisbury
M646EMR	Mercedes-Benz 1124F	Carmichael International	WrL/R	1995	4/4 Westbury
M647EMR	Mercedes-Benz 1124F	Emergency One	WrL/R	1995	1/6 Wooton Bassett
M648EMR	Mercedes Benz 1124F	Emergency One	WrL/R	1995	2/2 Corsham
M	Mercedes-Benz 1324	Carmichael International	WrL/R	1995	
M650EMR	Dennis Sabre TSD233	John Dennis	WrL/R	1995	2/1 Chippenham
	ERF EC10	GB Fire/	PHP	1995	4/1 Trowbridge
	Land Rover Defender 110 Tdi	Land Rover	L4V	1995	
N170KAM	ERF ES8	Multilift	PM	1995	1/4 Stratton St Margarets
N	Dennis Sabre TSD233	John Dennis	WrL/R	1996	
N	Dennis Sabre TSD233	John Dennis	WrL/R	1996	
N	Dennis Sabre TSD233	John Dennis	WrL/R	1996	
N	Dennis Sabre TSD233	John Dennis	WrL/R	1996	
N	Dennis Sabre TSD233	John Dennis	WrL/R	1996	

A512EPB was aquired in 1992. A955XOB was ex West Midlands FS in 1995. GHR618N ex ET. G522UEU was converted from a van in 1995.

ISBN 1 897990 52 9
Typeset by Bill Potter
Published by *British Bus Publishing*
The Vyne, 16 St Margaret's Drive, Wellington,
Telford, Shropshire, TF1 3PH
Fax: 01952 255669

Printed by Graphics & Print Ltd
Unit A13, Stafford Park 15
Telford, Shropshire, TF3 3BB